About the Author

Kevin J. Cox (B.A., Bates College, M.A., and Worcester State College) is currently the Science Department Head at Burncoat High School in Worcester, Massachusetts. Over the past twenty years Mr. Cox has taught A.P. Physics and Engineering. He is well known for his "hands-on" approach to teaching science. Mr. Cox is an avid supporter of FIRST Robotics and has served as the faculty advisor to the Team 1735 Green Reapers based at Burncoat High School. He is also a long time track coach at Burncoat.

About the Book

Did you know that Alexander Graham Bell beat Elisha Gray by just two hours in the patent of the telephone? …….. Did you know that Neil Armstrong was only fifteen seconds of fuel away from aborting the Apollo 11 moon landing? ……….Did you know that Alexander Fleming only discovered penicillin because he forgot to do the dishes before going away on vacation? ……..Did you know that Nicholas Copernicus didn't publish his sun-centered universe until he was on his deathbed?

These are just a few tidbits amongst hundreds buried in ***100 Great Moments of Science: A Look at Scientific Discovery from Archimedes to Einstein.*** Told in a concise format these events are uniquely understandable and garnered during 20 years of teaching science in the Worcester Public Schools. Whether you are a serious science student or just a history buff in search of a good book, Mr. Cox's award winning ***100 Great Moments of Science*** is a must on every book shelf.

This book is dedicated to my family, especially Kevin, Katie and Kelly Cox, and my extended family for sharing life's journey.

Special thanks to Deborah Cole whose loving support and tireless efforts in editing and shaping this book made it possible.

Acknowledgements

Special thanks to my brother Raymond P. Cox and my mother Irene Cox for their input. Thanks also to my father, Ray Cox Sr. who first told me many of these stories. Thanks to Kelly Cox for some editing. Thanks also to my physics classes, in particular, the Sedgewicks, Nick Smits, Ryan Hacker, Joseph Zanca, Robert Pratt, Taylor Blankenship, and Matt Allen. A tip of the hat also goes out to Colleen West, and Connor and Steve Cox.

100 GREAT MOMENTS OF SCIENCE

From Archimedes to Einstein and Beyond

KEVIN J. COX

Island Light Books

Copyright © 2009 by Kevin Cox
All rights reserved. Printed in the United States of America.

ISBN   1449542816

Cover Design by Ryan Hacker

Illustrations by Kevin Cox
    Selected Illustrations by Richard Ovian, – Crop Circles in Exo-planets, Hand of Sand in Galaxies

Clip art Courtesy of Dover Clip Art Series and Dover Pirtorial Archive Series Dover Publications, Inc., Mineola, New York, and

Clip Art Credits:
Charles Hogarth. Portraits of Famous People. Dover Clip Art Series.

Charles Hogarth. Famous Places.

Good Olde Days. Graphic Source Clip Art, Wheeling, IL @1986

Classic Comic Cuts. Graphic Source Clip Art, Wheeling, Illinois @1989

2000 Early Advertising Cuts, Dover Publishing, Clarence Horning @1956

                Printed. 01/18/10

Library of Congress Cataloging-in-Publication Data
Cox, Kevin J.
     100 Great Moments in Science. Describes 100 great moments in
      the history of science from Archimedes to Einstein and Beyond.
1.   Science – Biography – Juvenile Literature
2.   History – Juvenile Literature
10 9 8 7 6 5 4 3 2 1
EAN-13 9781449542818

Island Light Books

Contents

1. Man on the Moon 9
2. Man Flies 12
3. Anesthesia 13
4. Principia 16
5. Human Genome 18
6. Penicillin 19
7. Germ Theory 21
8. Evolution 22
9. Double Helix 24
10. Atomic Bomb? 25
11. Copernicus 27
12. Microscope 29
13. Lindbergh Flies The Atlantic 30
14. Light Bulb 32
15. Television 34
16. AC Power 36
17. Saturn has Ears! 38
18. Jenner 39
19. Mendeleev 42
20. Goddard 43
21. Dolly 46
22. Transistors 47
23. Telegraph 49
24. Telephone 50
25. Marconi's Miracle 52

26. And Yet It Moves 55
27. Kepler 56
28. Special Relativity 58
29. Tycho 59
30. Pendulums 60
31. General Relativity 61
32. The Trieste 62
33. Dinosaurs 64

34. The Miracle Year 66
35. Automobile 67
36. Sputnik 68
37. Mendel's Legacy 69
38. Big Bang 70
39. Franklin's Kite 71
40. Need a Lift? 73
41. Pangaea 74
42. Montgolfier Brothers 75
43. Tom Thumb 76
44. Pluto 78
45. Galaxies 79
46. The Curies 80
47. Camera Obscura 82
48. Dynamite 84
49. Under the Pole 85
50. North Pole 86

51. Black Holes 86
52. Edison's Favorite 87
53. Galileo's Freefall 88
54. Small Pox Ended 89
55. Finding the Titanic 90
56. A Most Unusual Experiment 92
57. Dwarf Planets 98
58. Brave heart 94
59. Bubble Wrap 95
60. Color Coding 96
61. Eureka! 97
62. Elusive Ether 98
63. Oxygen 99
64. Exo-planets 100
65. La Brea Tar Pits 102
66. Sound Barrier 103

67. The Dark Side   104
68. A Dog's Life   105
69. First Man in Space   105
70. The Engine   106
71. Archimedes' Ship   107
72. Hospital Standards   109
73. Voyager   110
74. Frog Power   111
75. Paperclip   112

76. Tesla's Death Ray   113
77. Discovery of Cells   115
78. Fission Vision   115
79. Germ Warfare   116
80. Quarks   118
81. Archimedes Death Ray   119
82. Deep Blue   120
83. Planet Vulcan?   121
84. Turtle   122

85. Charles Babbage   123
86. Roemer   124
87. DNA Fingerprinting   126
88. Hopeless Diamond   127
89. What Happened   128
to the Brontosaurus?
90. Halley's Comet   129
91. Space Elevator?   130
92. A Noble Gesture   131
93. Einstein's Compass   132
94. John Jeffries   132
95. Vineland Map   133
96. Wireless Power   134
97. Tunnel Vision   136
98. Bell's Other Invention   137
99. Serendipity   138
100. Earth Rise   139

1. Man on the Moon
10:56 p.m. (EDT) July 20, 1969

Arguably the greatest moment of science was the moon landing. Two Americans, Neil Armstrong and Edwin "Buzz" Aldrin, walked its surface on July 20, 1969. "If we can send a man to the moon" the saying goes, "we can do anything."

This greatest of all adventures began with a morning liftoff on July 16, 1969 from Cape Kennedy in Florida. It capped a decade of space exploration started by President Kennedy's challenge in 1962. "We choose to go to the moon, not because it is easy, but because it is hard" he said. On July 16th three brave NASA astronauts,

"That's one small step for man, One giant leap for mankind"– Neil Armstrong July 20. 1969

The Lunar Rover made its debut during Apollo 15

Neil Armstrong, Michael Collins and "Buzz" Aldrin blasted off atop the Apollo 11 rocket in a 4 day, 239,000 mile jaunt to the moon.

On July 20, 1969, they had reached the moon and were ready to descend. An anxious television audience of over a billion people huddled around their black and white television sets to see the event. Astronaut Michael Collins would circle the moon with the command module Columbia, while "Buzz" Aldrin and Neil Armstrong descended to the lunar surface. Only a few layers of thin tin foil protected their Lunar Excursion Module (LEM) from the crushing void of space.

Back at NASA's Mission Control, Flight Director Gene Kranz paced the console room where one hundred or so engineers monitored the flight's progress. Then trouble began, the tiny computer aboard the LEM overloaded. It had a little more brainpower than our modern handheld calculators of today. It was also clear they had overshot their landing site and were headed for a crater. They were also dangerously low on fuel. A

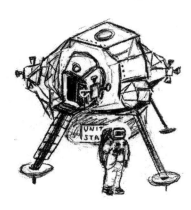

The Lunar Excursion Module (LEM) was said to resemble a spider.

quick thinking Armstrong switched to manual control of the module. He throttled the LEM forward with only about 30 seconds of fuel left. A large stopwatch in the console room at NASA carefully chronicled the time left.

"700 feet, …. 400 feet, …. 75 feet" chronicled the 38 year old pilot. Both Aldrin and Armstrong were standing since the craft was so small there was not enough room to sit. It was no bigger than a phone booth. Aldrin strained to look out their small triangular windows. Finally Armstrong declared "dust kicking up," a sure sign that they were within a couple of feet from touchdown.

"Tranquility Base here, the Eagle has landed." radioed Armstrong.

"We've got a lot people breathing again," responded Krantz. There was less than fifteen seconds of fuel remaining when the LEM landed. High-fives and clapping filled the room. It was July 20, 1969 at 4:17 p.m. (EDT). No matter what else happened from here, at least they had made it onto the surface of the moon.

Six hours later, Armstrong and Aldrin were ready to walk the surface. They donned their bulky 300-pound spacesuits, which only weighed 50 lbs on the moon. Armstrong opened the hatch and reported, "Houston, I'm on the porch." A camera at the foot of the LEM captured Armstrong's black silhouette against the stark white background.

He slowly descended the LEM ladder. The ladder looked more like it belonged to a tree house than a six billion dollar spacecraft. His giant boots and bulky spacesuit made him look more like an arctic explorer than a visitor to another world. He noted that the last step was a bit steeper as he hopped onto the final footpad. At 10:56 p.m. (EDT) Armstrong stepped out onto the lunar surface and declared, "That's one small step for man, one giant leap for mankind."

Next, "Buzz" Aldrin emerged. He noted to Armstrong that the door would not be locked, a detail Armstrong appreciated hearing. After descending the ladder and walking a bit, Aldrin began leaping about the surface to play in the $1/6^{th}$ gravity and to better understand lunar locomotion. Aldrin described the moon as a "magnificent desolation."

In all, they walked the surface for a mere two and a half hours. Later missions would stay much longer including Apollo 17 which stayed for three days in 1972. On this day the two astronauts would plant the American flag, place a plaque, and take a phone call from then President Nixon. They collected some forty-seven pounds of lunar rocks, set up an additional camera and then retired for some much needed rest.

Shortly after 11 a.m. the next day Armstrong and Aldrin prepared for blastoff. Only the top half of the LEM would take off, the bottom would stay on the moon and act as a launch pad. Later they would rendezvous with the orbiting Michael Collins and begin their three day journey home. A successful splash down in the Pacific Ocean ended mankind's greatest adventure.

Six other Apollo missions would follow. In all twelve men walked the moon; the last was Apollo 17 in 1972. Each time man would stay longer and do more. Apollo 15, 16, and 17 even sported a car, which extended their exploration range up to twenty miles at an 8 miles per hour clip. Armstrong noted that the moon was covered with a very fine powder about an inch thick.

When will we be back? It's been nearly fifty years since Apollo 11. The discovery of helium-3 on the moon provides some financial impetus to return. Helium-3 is a rich fuel used in fusion which is extremely abundant on the moon but almost non-existent on earth. It has the potential to make oil and nuclear fuel obsolete. The hope is that this will spur a second race to the moon. A permanent manned moon station is planned for the year 2020, but it is always subject to budget cuts. It is planned for the South Pole of the moon, which is always visible to earth. Until then we must wait and watch. Because there is no weather on the moon, Armstrong's footprints will remain for thousands of years if not longer. We had walked another world and in so doing forever changed the one we left behind.

The moment of flight was captured
By John T Daniels

2. Man Flies
The Wright Brothers Achieve Flight - Dec 17, 1903

Two brothers, Orville and Wilbur Wright, thought they had cracked the riddle that had befuddled the greatest minds of the millennia, flight. On December 17, 1903 they came to the dunes of Kitty Hawk, North Carolina, to challenge the demon, gravity, that had shackled mankind to the ground since time began.

A stiff cold wind blew over the biting sands as Orville and Wilbur Wright shivered in their starched shirts and customary business suits. Orville hung out the signal flag to alert the locals, mostly fishermen, to help them maneuver their flying machine onto the dunes.

With a flip of the coin, lady luck put the younger, thirty-two year old Orville, into the pilot's position. There was no seat; instead the pilot had to lie, stomach down, in the center of the plane. He would then have to use his body weight to help shift control of the plane. There were also ropes and pulleys to control its flight. They invented an ingenious method called wing-warping to twist the wings back and forth and provide control. It was in this area of control that the Wrights were far in front of the rest of the world. While others like Ernest Dumont of France could get their planes up in the air they could not bank and turn as the Wrights. Control, that was the key, something they had learned at their bike shop in Dayton, Ohio.

Orville positioned himself belly down on the lower wing of their double-winged craft. Their self-built engine sputtered and cracked in the crisp December air. Orville released a small lever and the machine lurched forward riding atop a short rail set on the dunes. Then it happened, the Wright flyer lifted into the air. Wilbur gasped, with arms out and eyes wide open. A shocked photographer John T. Daniels snapped a picture of this moment. It was 10:35 a.m. Twelve seconds and 120 feet later the flyer came back down to a soft-cushioned stop. They had done it! The Wright Brothers had flown.

They would make three more flights that day the longest being fifty-seven seconds and 852 feet. After the final flight an errant gust of wind sent the idle flyer cart wheeling, damaging it forever, but it didn't matter. Wilbur dashed off a quick telegram to their minister father Milton back in Dayton, Ohio.

Wilbur Wright (1867-1912)

Orville Wright (1871-1948)

In the next few years, the Wright brothers continued their advances. They added a dropping weight and a catapult which eliminated the need for the constant winds. With this they were able to move operations back to Dayton, Ohio at Hoffman's prairie. By 1908 they were making flights of nearly seventy minutes and twenty-four miles.

Many were incredulous about the Wright brothers claim. The Wrights were very secretive and refused to demonstrate their flyer until their patent came through. Finally in Paris on August 8, 1908 the Wrights demonstrated their machine and all that were in attendance were amazed. Clearly they had succeeded far beyond anyone to date.

The Wright Brothers' story is one of the greatest in history. Self-funded, outsiders, armed with only a can-do attitude and an engineering brilliance, they achieved what no one else in history was able to do, that is, the impossible.

3. Anesthesia
Morton Demonstrates Anesthesia
October 16, 1846

In the days before anesthesia the word "operation" would strike terror into the most stoic of patient's hearts. It was a gruesome experience for both the doctors and the patient. If you were wounded in battle you might be brought to a field hospital where several strong men would hold you down and a good strong doctor would do his best to sever your leg in less than 60 seconds. But on October 16, 1846 this hellish reality began to recede as the days of anesthesia and painless surgery began. Eventually the doctor's crude saw was replaced by the prick of a needle and the nightmare became a deep sleep instead. Anesthesia, operations without pain, is one of the single greatest medical discoveries of the last millennium.

The story of anesthesia has its roots in college dormitories where medical students would breathe ether or nitrous oxide for kicks. These "ether" parties soon turned into laughing jags, or giddiness and revelry. Because of this, nitrous oxide was nicknamed "laughing gas." It was noted by many that if one injured one's self during this activity, one seemed not to notice it.

On December 10, 1844 after seeing a public demonstration of laughing gas, Boston Dentist Horace Wells had the idea to use it while extracting teeth. One of the demonstration participants, Samuel Cooley, had gashed his leg and was bleeding profusely. Wells noted that Cooley was amazingly unaware of his injury while under the effects of the nitrous oxide.

The next day, December 11, 1844 Horace Wells asked Dr. Riggs to come to his office, administer the laughing gas, and extract his tooth. It was a great success. After successfully extracting a dozen or so patients with

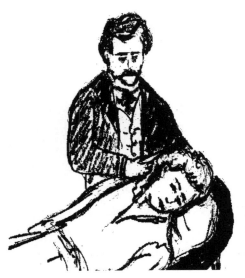

nitrous oxide, Wells set up a demonstration at Harvard University. Unfortunately, it did not go well. The patient screamed with pain as he yanked on his molar and the onlookers jeered and booed. His reputation was ruined and he eventually took his own life in 1848.

Wells' former dental partner, William T. Morton took up the cause. Instead of nitrous oxide however, he used ether which would produce a much deeper sleep. First trying it out on insects, he then moved on to goldfish, dogs and then himself. After two or three years of experimentation and success, he was ready for his first public demonstration of his anesthesia. The failure of Horace Wells was very much in his mind but he forged ahead.

On October 16, 1846 William Morton gave the first successful public demonstration of anesthesia during an operation. The setting was the operating room of Massachusetts General Hospital in Boston. The patient was a printer named Gilbert Abbot. A full crowd of distinguished doctors were on hand to watch Dr. John Collins Warren remove a tumor form Mr. Abbot's neck.

William T. Morton, a dentist by trade, slowly jogged the three blocks from his laboratory to the Massachusetts General Hospital. In his pocket was the magic elixir he had dubbed "Letheon" which would hopefully usher in painless surgery. It was really just ether with perfume added.

He barged into the side door where a waiting Dr. John Warren looked quite perturbed. Morton took a large sponge, soaked it in the magic "Letheon" and placed it into the inhaler jar. Mr. Abbot was told to relax, a tall order considering the circus swirling about him.

Mr. Abbot breathed deeply from the inhaler as an assistant pinched his nose. Within three minutes Abbot was asleep. A slight snore gave Morton the confirmation he needed.

"The patient is yours doctor," Morton confidently informed Warren. A collective sigh let out from the crowd as he cut into Gilbert's neck. Blood began to pour out, but no screams. The silence in the room was deafening. A look of shock came over the small crowd. Even the 70 year old Dr. Warren paused momentarily in disbelief before dispatching his operation. A groggy Abbot regained consciousness; his arms and legs feeling heavy as tree trunks.

Dr. Warren leaned in and quietly asked "Did you experience any pain?"

"No," replied Abbot. It was a miracle.

"Gentleman," Dr. Warren rang out, "This is no humbug!" The assembly broke into applause. The gravity of these events was felt by all, especially Abbot. Two months later, on Dec. 21, 1846, an even more dramatic demonstration would take place in London. Dr. Robert amputated a complete leg from Frederick Churchill, a 36 year old butler, again without pain.

The era of painless surgery had begun. The exact agent of anesthesia would change over the years but the breakthrough had been established. The crude saw of the army surgeon was eventually replaced by the gentle prick of a small needle, and for that we can all sleep a little easier.

4. *Principia*
Newton Writes the Definitive Work on Physics
1687

What Mozart is to music, Sir Isaac Newton is to science. His book has been the cornerstone for classical physics for over 200 years. Time magazine in 2000 named Newton the most influential person of the last millennium. That our planes fly and cars run are due in no small measure to his contributions. It was Galileo (1564-1642) who mathematically described *how* objects move, *kinematics*, and Newton in his *Principia* of 1687 described *why* objects move as they do.

**Sir Isaac Newton
1643-1727**

Newton was the genius to the geniuses, gifted with one of the highest IQs ever estimated at 190. *Principia* answered some of the most profound questions of science, in particular on motion and astronomy. In it, Newton states quite simply his three laws of motion. First, a body in motion stays in motion, unless acted upon by an outside external force. A body at rest stays at rest, unless it too is acted upon by an outside external force. Second, the acceleration of an object is proportional to the force acting upon it. Third, for every action there is an equal but opposite reaction. Forces are like socks, they come in pairs. He then added his law of gravitation to explain the falling of an apple and the orbit of the moon and planets.

Newton conceived most of the ideas for *Principia* when he was sent home from London after an outbreak of the plague when he was 24. During this eighteen months starting in 1666 he also invented Calculus, called "Flexions." He would not publish his works for another twenty years until 1687. He would also make major breakthroughs on the theory of light.

Sir Isaac Newton tells the story that on one morning in 1665, while reading under his favorite tree at Woolsthorpe an apple fell to the ground. At this moment he supposedly realized the unifying principle of gravitation. He understood that the moon, like the apple, was also falling towards the earth but never hitting it. The earth curves out from under it faster than it falls. Newton eventually shelved the project for twenty years because he couldn't get the mathematics to agree. Unbeknownst to him at the time, he did not have the correct distances of the moon's orbit.

Twenty years passed before Sir Edmund Halley knocked on his door in 1684. The affable Halley, also known for the discovery of Halley's Comet, told Newton of a fireside chat that he once had with Sir Christopher Wren and Robert Hooke. He asked Newton, "If a planet were to follow an inverse square law such as gravity what shape would the orbit be?"

"Why an ellipse," Newton quickly responded. A white faced Halley froze in his seat.

"Are you sure? This is a miracle. This is exactly the orbit that Kepler had predicted!" Halley could scarcely believe Newton's answer. An excited Halley asked Newton for the mathematical proof. Newton rummaged about pretending to find it and then replied that he would have to send it later.

In fact, Newton had not finished the work he had started twenty years earlier. For the next two years, Newton worked feverishly and finally published *Principia* in 1687. "If I have seen further," Newton says, "It is because I have stood on the shoulders of giants." He died in 1727. On his monument reads a quote from Alexander Pope, "Nature and Nature's laws lay hidden in the night, God said 'Let Newton be,' and all was light." John Squire would later add in response "It did not last; the Devil howling 'Ho! Let Einstein Be' and restored the status quo."

A genius among geniuses, Sir Isaac Newton wrote most of his important works at age 24 in the eighteen months he stayed at his home of Woolsthorpe, England in 1666. London was in the grips of a horrible plague.

5. The Human Genome Project
June 26, 2000

The mapping of the Human Genome, completed in 2003, has been called by President Clinton "the most important, most wondrous map ever produced by humankind." It is quite simply the instructions for how to build and maintain a human and understand its functions. The human DNA is a set of 3,300 million instructions known, as based-pairs, stored into 23 chromosomes inside each cell. These instructions control everything from the initial building of the human being to its repair and daily metabolism throughout a person's life.

If each based-pair were a printed letter on a page of a book, the complete instructions would fill nearly 11,000 volumes and reach 1,650 meters high. That is over 5 times the height of the Empire State Building. If they were actual rungs of an actual ladder, since the DNA has been described as a twisted ladder. That ladder would stretch from the earth to the moon and back again.

Despite its enormity, this DNA ladder has been climbed. At a cost of $10 billion and 15 years of work thousands of scientists at Celera Inc. and Government facilities have completed this task equal in scope and difficulty to the landing of a man on the moon. It is one of the crowning achievements of 20th-century science and it opens up a whole new branch of medicine. As President Clinton's Press Release announcement of June 26, 2000 aptly says it "will lead to new ways to prevent, diagnose, treat, and cure disease. Alterations in our genes are responsible for an estimated 5000 clearly hereditary diseases, such as Huntington's, cystic fibrosis, and sickle cell anemia, and influence the development of thousands of other diseases."

A key finding of the Human Genome Project included a discovery that only a small percentage of these based-pairs include genetic information, 24,000 genes out of 3,300 million pairs, that is 1 thousandth of 1 percent. The second is that we are all 99.99 % identically alike in DNA. President Clinton, joking that he was getting in touch with his "inner nerd" noted that that one-tenth of 1 percent difference between he and the esteemed scientists is the capacity to do science.

Whether the Human Genome Project fulfills its hype, only history will tell, but its potential is enormous. It is a bit like asking "What good is electricity or computers?" when they were still in their preliminary stages. The applications of the Human Genome Project are endless and like electricity and computers, it will transform our world. Hopefully with this map humanity can chart a better course health wise and make our life's journey a bit less periless.

6. Penicillin
Alexander Fleming Discovers Penicillin
August 1928

Dr. Alexander Fleming (1881-1955)

In the days before antibiotics death was as near as a pin prick from a rose petal. This happened to one unfortunate policeman, Albert Alexander, in 1941. A bacterial infection set in, and there was no treatment to stop the ravages of the disease from taking his life. He was one of countless millions that lost their lives to bacterial infection.

Almost every household had lost at least one member of its family to bacterial infection. It was particularly hard on children where one in four children would die before their first birthday. Even the indomitable George Washington died of infection when he contracted strep throat in 1797.

In 1928, English doctor Alexander Fleming discovered penicillin by accident. It turned out to be the miracle drug of the century, and ushered in the new weapon of antibiotics for mankind. He first found, and then lost the cure, only to have it found again by Ernest Chain ten years later.

In August of 1928 Dr. Alexander Fleming (1881-1955) and his family were packing for their annual two-week summer vacation outside of London. As usual Fleming had too much to do before he left and he stacked a number of unwashed culture plates on his laboratory bench at St. Mary's Hospital. They were filled with a particularly deadly form of bacterium called staphylococci.

After a restful vacation Fleming returned two weeks later and began to clean up where he left off. As he began to deal with his unwashed culture plates he noticed that the bacteria were completely missing adjacent to a piece of mold in the plates. The cell walls were broken down leaving just a shell of the bacteria. As one moved away from the mold the bacteria gradually became fuller and more numerous. He identified the mold as penicillin, a very common household mold. Fleming correctly surmised that it was the penicillin mold that had killed the bacteria close to it.

This should have been the end of a great story that reads everyone lives happily ever after but it is not. The miracle drug penicillin would need a second hero to bring to the masses. In eighteen months, Fleming was unable to isolate the active ingredient in the penicillin mold. Funding requests to St. Mary's for further study were turned down. In addition, the penicillin seemed to lose its punch after just a few hours. A disillusioned Fleming published a paper on it, and then the greatest medical discovery of the twentieth century was shelved for ten years completely unnoticed.

In 1938 Chemist Ernst Chain rediscovered Fleming's papers in 1938. Chain managed to do what Fleming could not, isolate the pure ingredient penicillin from the mold. Assisted by his partner Howard Florey, Chain and Florey tested the compound on six very ill pediatric patients with miraculous results. All six patients survived.

Mass production of penicillin began by Merck, Pfizer and other chemical companies and by 1945 enough penicillin was produced to treat 45,000 cases. For their work Fleming, Chain and Florey shared the Nobel Prize for medicine in 1945. As noted in a 2004 Modern Marvels episode on antibiotics, "It has been estimated that Fleming's discovery and subsequent antibiotics have added ten years to the life expectancy to every man, woman and child on the planet." It opened the door for other antibiotics and soon we had nearly a dozen weapons to repel bacterial invaders.

Today bacteria are mounting a counterattack. Bacteria are changing, mutating and becoming drug resistant. Doctors are beginning to see this dark spector of infection rising again. Today medical scientists are working feverishly to find new antibiotics for new and old enemies. A return to the day of pre-penicillin where infection ran rampant is an unacceptable future, a bitter pill that none of us wish to swallow.

7. Germ Theory
Louis Pasteur Develops Germ Theory
1870

"Achoo!" someone sneezes. "Guzenthiet," we say or "God Bless you!" It's a tradition dating back centuries, when we believed that illness might be caused by evil spirits or perhaps God's displeasure. During the plague of the middle ages, doctors would wear fearsome masks to scare off evil spirits.

That all changed with Parisian Louis Pasteur in the 1870's. "Microbes are the culprits," he screamed. These enemy microbes are too small to be seen with the naked eye. They are what is making us sick. It is also bacteria in milk and wine that causes it to sour. By heating the milk first, called "pasteurization," the milk would stay fresher much longer. It should also be noted that not all bacteria are bad. Good bacteria in yeast cells make wine ferment, and the benefits of yogurt are well known.

**Louis Pasteur
(1822-1895)**

Sickness was caused by bacteria from outside the body.

Pasteur not only focused medicine on the causes of sickness, he also preached the importance of teaching the body to mobilize its defenses against these diseases. Like Edward Jenner with small pox, Pasteur pioneered a vaccine for rabies and anthrax.

Pasteur's work was the beginning of modern bacteriology and set the course of medicine for the next 100 years. It would be left for others to search out and destroy these pathogens and restore us to health. Finally, we had an understanding of the causes of diseases, which is the first step toward curing them. It was truly a great moment of science and an accomplishment not to be sneezed at.

8. The Evolution Revolution
Darwin
1859

In 1836 Captain Fitzroy of the HMS Beagle had a lot on his mind, a long trek across the Pacific, a treacherous turn around the horn of Africa, and then the icy Atlantic before returning home to his native Plymouth, England. Yet the most anguished man on the boat was not Captain Fitzroy, but a twenty-two year old botanist Charles Darwin. He too was headed for treacherous waters, walking a line between religion and science, a battle he wished not to face. The ideas he was tossing and turning in his head would fly in the face of accepted religious doctrine. It was his theory of evolution.

As the H.M.S. Beagle left the Galapagos Islands, 600 miles west of Ecuador, naturalist Charles Darwin crated up his innumerable specimens along with his bible, a hammer and a gun with one bullet. There were dozens of species of birds and beetles along with his copious notes. Captain Robert Fitzroy barked a few final commands from the deck of his 90-foot by 8-foot vessel.

**Charles Darwin
(1809-1882)**

"Steady as she goes" he bellowed "Let her breathe."

It was the end of a five-year mission that included Brazil, Argentina, Chile, Peru and the Galapagos Islands. Isolated from South America, the Galapagos Islands are called "the island that time forgot." There were species found here that were unlike any in the world such as birds that swim and huge tortoises one could straddle and ride on their backs. For five weeks Darwin explored these unique islands, collecting a huge amount of specimens. Of particular interest were the different kinds of finches found on the islands. It was clear that these finches were all closely related, but there were differences in their bill structure.

Darwin reasoned that the scarcity of the food which differed from island to island caused some finches to adapt to survive better than others. If the predominant finch food was big seeds, the finches found on that island had strong beaks for breaking big seeds. If the predominate food was cacti, finches with long pointed beaks became predominant. A third finch, the woodpecker finch, had a beak exceptionally good for grasping insects and knocking on the trees to bring the hiding insects to the surface. Did God make a new species for each island or did they have a common ancestor and then adapted? Species "adapt" or change to survive in their environment. Those adaptations which help the finches survive became more numerous while the others died out. They became the dominant finch on that island. Darwin called this the "survival of the fittest."

It would take Darwin twenty years to publish his ideas. His book on evolution, officially titled the *Origin of the Species* was finally published in 1859. It had taken over 20 years to complete. When Darwin finally did publish it, Christian church officials were outraged. It flew directly in the face of instant creation in six days as described in the Bible. It taught that the millions of species on the earth were all created in their final form at the same time and zapped onto a waiting earth. Many creationists argue that an infinite amount of variations would be needed to produce the complexity of the human eye. They would also say that the fossil record is incomplete with whole transition periods missing.

It should be noted that Darwin was not anxious to tangle with religion. In his later life he seemed to have the weight of world on his shoulders, including depression, stomach ailments, sleeplessness, and heart palpitations. Having once been a Divinity student he was tortured by a fear of eternal damnation. Even his wife Emma pained to think of the fate of his soul and told him so. For Darwin, he simply believed evolution to be true. He didn't have a choice any more than Galileo could deny that the earth moved. If evolution could have been rationally shown to be false nobody would have been happier than Charles Darwin to finally get the "monkey" off his back, so to speak. For those that accept his theory, a lot of things make sense. For those that reject evolution for religious or other reasons, Darwin would preach tolerance, after all variations within a species are to be expected.

9. The Double Helix
Watson and Crick Find Structure of DNA
1953

It was known since Gregory Mendel's work in 1860 that hereditary information is passed from parents to offspring via DNA. In 1902 Danish biologist Wilhelm Johannesen first used the word "gene" as the carrier of this heredity information. Walter Sutton in 1909 correctly placed these genes on the chromosomes found inside every cell. Thomas Morgan later showed that this information was found to be carried in the 46 human chromosomes. The exact structure of this DNA molecule was still very much a mystery. What does it look like? What is it made of?

The riddle was finally solved by two doctors in 1953, English Dr. Francis Crick (1916-) and American Dr. James Watson (1928-) and the X-ray work of Rosalind Franklin. Working in London, Watson and Crick built a six-foot three-dimensional model of DNA out of clamps, steel wire, and tubing. They had gotten key insights by examining the x-ray photographs provided by Rosalind Franklin.

The structure turned out to be surprisingly simple and elegant. It was a twisted ladder, a geometric shape known as a double helix, with each rung of the ladder representing an instruction for building amino acids. The instructions were written in only four letters A, C, G, and T which stood for adenine, cytosine, guanine and thymine. Adenine only combined with thymine while guanine restricted itself to cytosine. These base-pairs formed one rung of this 3.3 billion rung ladder.

The massive job of detailing this 3.3 billion rung ladder would not happen until 2000 in the mapping of the Human Genome Project, but

Watson and Crick's 1953 breakthrough is considered one of the most important moments in scientific history. They received the Nobel Prize in Medicine in 1962 for their work. Franklin was ineligible as the Nobel Prize is not given posthumously and Franklin died of cancer in 1958.

10. The Atomic Bomb?
July 16, 1945

Robert Oppenheimer (1904-1967)

A modern sword of Damocles hung from the top of a 100 foot steel tower. It was the morning of July 16, 1945, in a remote New Mexico location called Jornada del Muerto, Spanish for "Journey of the Dead." It is here that a dedicated group of American scientists, led by Robert Oppenheimer, gathered to detonate the world's first atomic bomb. Thirty-eight year old Robert Oppenheimer waits nervously in a concrete bunker six miles away. In the balance is not only the bomb, but the balance of power in World War II. A successful explosion could be the knockout punch to end the brutal war with Japan.

Oppenheimer and his team lowered their thick welding glasses. The intense blast will be one hundred times brighter than the sun. At 5:30 a.m. the five-foot ball dropped. A thundering rumble violently shook the earth. Houses blew apart as if built by playing cards in the path of a vacuum cleaner. A great burst of light swallowed the sky as telephone wires buckled and bent under the tremendous shock harrowing its way. The heat vaporized everything in its path as it reached over a million degrees. Steel burnt like wood. Its now familiar mushroom cloud sucked every bit of dust five miles high before curling outward. Surely even Armageddon could not be worse.

"Oh my God," Oppenheimer mumbled. Others could not speak. Some laughed, some cried, but most were just stunned. General Groves telephoned President Truman to inform that his new weapon was ready. Would he use it?

Truman issues a last ultimatum, which the militant Japanese flatly refused. The order was given and the new super weapon was secretly flown in pieces to Tinan Island some 1,600 miles east of the Japanese mainland. On August 5, 1945 a B-29 bomber, the Enola Gay, was specially outfitted and the bomb, nicknamed "Little Boy," was loaded into the cargo area. The plane was so heavy a nervous Captain Tibbits wondered if the plane would ever get off the ground. He rumbled down the long runway and at 2:45 a.m. the Enola Gay was airborne. Six thousand miles away, in Los Alamos New Mexico, Oppenheimer awaited the news.

At 8:15 a.m. on August 6, 1945 "Little Boy" made its rendezvous with destiny. It was released from the cargo bay and fell screaming towards an oblivious city of Hiroshima enjoying a beautiful day. Children were laughing and riding bicycles. People were scurrying to work not knowing that it would not matter sixty seconds later. A stressed Captain Tibbits banked hard to escape the coming fury.

Navigator "Dutch" van Kirk witnesses a blinding light as the bomb struck home. "Oh my God! Oh my God!" he mumbled unable to turn away. A violent shock wave soon overtook the plane and shook it helplessly in midair, but it flew on. In New Mexico it was only desert and dirt, but today it was flesh and blood. It was women, children and the second largest Japanese army of 43,000 soldiers strong. It also held large munitions and had shipyards of strategic military value. Seventy thousand people would lose their lives instantly while many thousands would later die of radiation poisoning from the blast. A four-mile radius of the city was obliterated. It was a horrible end to a brutal war.

Five days later, a second bomb, "Fat man," was dropped on Nagasaki with similar results. A shocked Japanese leadership finally relented. Emperor Hirohito asked his country to "bear the unbearable" and surrendered aboard the USS Missouri, ending WWII.

Albert Einstein once said "I know not with what weapons World War III will be fought, but World War IV will be fought with sticks and stones." The specter of nuclear holocaust haunts us daily. Years later, a glassy eyed Oppenheimer regretted his part in the bomb. Reminiscent of Alfred Nobel's sleepless nights, Oppenheimer recalls a line from the Hindu scripture, "I am Vishnu destroyer of worlds."

Was it a great moment of science? It was certainly an important moment of science and history that changed life on earth forever.

11. Copernicus
Copernicus Puts the Sun in the Center
May 24, 1543

On May 24, 1543 seventy year old Nicolas Copernicus lay dying in a church tower in Poland. The first copy of his historic book, *Concerning the Revolutions of the Heavenly Spheres*, was pressed into his near lifeless hand. It had been forty years in the writing and Copernicus was too weak to hold it.

In his work Copernicus overturned the earth-centered solar system accepted for thousands of years and instead put the sun in the middle. The earth was simply another planet orbiting the sun just like Mars, and Venus.

It was a model that the Catholic Church vehemently opposed. They would even burn Father Bruno at the stake in 1600 for simply professing this view. Since man is the center of God's plan of creation, the church reasoned, God would not have relegated the earth to a secondary position subordinate to any other celestial body. They defended the second century Ptolemy view that the earth, not the sun, was the center of the solar system, and the other celestial bodies, including the sun orbited around it.

Nicolas Copernicus
1473-1543

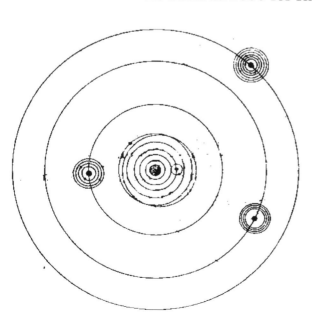

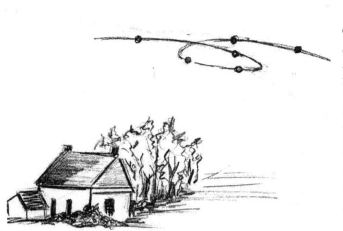

Some planets appear to move backwards over a few nights observation

But the old earth-centered universe simply just did not fit the data, in particular the motion of Mars, Mercury and Venus. There are times in the years when these planets appeared to move backwards. It is known as retrograde motion. This can easily be explained by the Copernicus sun-centered model. The earth, having a smaller orbit than mars, for example, overtakes mars in its orbit around the sun. This makes mars appear to move backwards for a few days. It is much the same as when one passes another car on the highway. The other car appears to move backward when in fact both of the cars are moving forward.

The Ptolemy system could never successfully explain this retrograde motion. Instead they employed extra circles, called epicycles, to explain the retrograde motion of Mars. Mars would travel along its normal orbit, and then take a side circle still at different points of its motion. The number of side-circles needed to accurately describe the observed planets had risen to over 1,000 by Copernicus' time! It was obviously wrong.

Copernicus tried to put a first page disclaimer on the front, "For Mathematicians Only." It was marketed as a "what-if" model only not to be accepted as a "real" universe. This thinly veiled attempt fooled no one; least of all the Church and the book was banned in 1616. Private copies circulated all over Europe and many in the scientific community accepted it anyway.

Copernicus would die just a few days after the first printing of his books. The ban on his book would not be officially lifted until 1835. The fight for the Copernican model would be left to another, Galileo Galilei, would pay a terrible price for his infidelity. Copernicus' system was not without error. Copernicus had planets moving in perfect circles which we now know are ellipses. Their speeds are not uniform but rather they speed up as they approach the sun and slow down at great distances away. Still, it was a significant improvement over the Ptolemy system. Someone had to start the earth moving, and that someone was Nicolas Copernicus.

12. The Microscope
Leeuwenhoek Invents the Microscope
1677

Anton Van Leeuwenhoek (1632-1723)

In 1663 Anton van Leeuwenhoek (1632-1723) was leading a double life. By day he maintained a dry-goods store in Delft, Holland but at night he was Anton the explorer. Anton van Leeuwenhoek was the inventor of the microscope. While Galileo pierced the heavens for "ears of Saturn," in 1660 Anton was busy with the silent marvelous world at the other end of the meter stick, the world of the very small.

It all began innocent enough, something to take the drudgery out of an otherwise mundane existence. At night he would grind by hand the most exquisite lens known to man, a process he would keep secret his entire life of ninety years. Before he was done he had hand ground 247 lenses of which only 7 survive to this day. Each was a handheld device not much bigger than a Popsicle stick. It had a tiny one-eighth inch spherical lens, much like a peep hole in a hotel door. It also had a pointed stick on which to place the object to be at the focal length and two screws to move it slightly about.

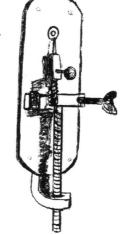

Armed with his new device human hairs suddenly became as thick as a man's arm. Flies became prehistoric beasts beneath his magnifying lens. In 1677 Leeuwenhoek placed a drop of pond water under his seeing eye and to his amazement found a whole variety of sea-creatures. Little water animals that swam and cavorted with internal structures to show they were not just randomly moving particles. They seemed to have purpose and navigation. He called these little creatures "animalcules." He was the first to see bacteria. At his best, Leeuwenhoek was able to magnify up to 300 times an object's original size. Today thanks to the electron microscope invented in 1931 by Ernst Ruska we are able to see up to 1 million times an object's original size. At these powers of magnification a postage stamp would grow to about 13 miles by 13 miles, or about the size of a small town.

It would be two hundred years before Pasteur would tie deadly illnesses to these innocent looking creatures but the enemy had been sighted. The microscope would be an essential tool in fighting this enemy and it was van Leeuwenhoek, a shop keeper from Delft, that first opened the window to an unseen world.

13. Lindbergh Flies the Atlantic
August 5, 1927

On August 5, 1927, twenty-five year old Charles Lindbergh was attempting to be the first pilot to cross the Atlantic. It was a feat that had already claimed the lives of several other aviators.

If only he had gotten just a few hours of sleep while waiting for the several days of constant rain to stop. His window of opportunity finally opened at 7:52 a.m. Lindbergh and his assistants rolled his silver plane, the Spirit of St. Louis, out on the one-mile runway at Roosevelt Field. The airport was located just outside of New York City.

Lindbergh put one hand on his single-engine plane and with his eyes inspected the rain-drenched runways. Camera bulbs popped nonstop as a multitude of spectators and reporters were here to either witness history or to be the last people to see him alive. Many called it a suicide mission. Lloyds of London, who will bet on anything, refused to bet on this trip.

Charles Lindbergh (1902-1974)

"Good luck, Lindy!" one said. He would later acquire the nickname "Lucky Lindy" but for now his friends called him "Slim." Perhaps this was due to his slender build. He was 6 foot 3 inches in height. With 425 gallons of gas, the Spirit of St. Louis was anything but slim. It was burdened to the max. In fact the gas tank was so large that it blocked the front window.

Lindbergh would have to fly much like a submarine, he could only look forward by a series of mirrors.

In front of him lay an ominous set of telephone wires and trees which he would need to clear at the end of the runway. Lindbergh climbed inside to his light wicker seat. Everything in his plane was built with conservation of fuel in mind. He did have with him a small inflatable raft, a three-gallon thermos, a ball of string, and a knife.

His assistants spun the propeller starting its Whirlwind engine sputtering at 7:40 a.m. The Spirit of St. Louis hummed down the runaway and rode the air, just barely clearing the telephone lines by twenty feet. "Next stop Paris, Le Bourget Field," he thought, some thirty-six hours east. He would average about 100 miles per hour during the trip.

At 9:12 a.m. Lindbergh passed Boston and the tip of Cape Cod, Provincetown. He carefully plotted his position on a large map, which he would do faithfully each hour. By 8:15 p.m. Lindbergh was over his last bit of land, St. Johns in Newfoundland and then darkness as night fell. It was a moonless night which would add to the crushing darkness. He would have to fly low at times, just a few hundred meters above the waves to see the white crests pointing the way.

By the 16th hour, 11:00 p.m., the fatigue and the night engulfed the 9,000-pound craft. Ice was beginning to build up on his wings.

Lindbergh buzzes a herd of sheep over Dingle Bay, Ireland in 1927

Both his eyes and his wings were heavy, but Lindbergh fought through the micro-sleep. He couldn't sleep while flying because every hour or so he had to balance off the fuel on each side of his plane. At one point he thought he saw a great sailing vessel but it was just a large iceberg. In fact there were lots of icebergs.

On the 28th hour Lindbergh's eyes were like slits. His bad posture could hardly be described as sitting. He tried raising his eyebrows several times but his eyelids refused and his head kept crashing to his chest. Finally, a seagull came into view. "Seagulls don't stray too far from land," he reasoned. He figured that he must be close. Within an hour he saw fishing boats and the coast of Ireland. It was 10:52 a.m. Lindbergh was giddy with excitement. He recognized the row houses and correctly surmised that it was the fishing village of Dingle Bay on Ireland's southwest coast. A rejuvenated Lindbergh buzzed a herd of sheep grazing in the Irish countryside.

"Only three more hours to England, Paris is still another 600 miles but getting closer," he thought.

Finally at 10:22 p.m. New York time, thirty-three and a half hours and 3,610 miles, Paris! Lindbergh circled the Eiffel Tower and brought his plane, the Spirit of St. Louis, down on Le Bourget Field. A throng of 10,000 waited for him and overran the barricades. Lindbergh was afraid his moving propeller might hurt someone and brought his plane to an early stop. Pandemonium erupted! For nearly half an hour Lindbergh was lifted aloft unable to touch the ground. The spectators ripped off parts from his plane. A quick thinking Frenchman put Lindbergh's helmet on a reporter and yelled, 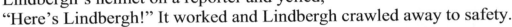 "Here's Lindbergh!" It worked and Lindbergh crawled away to safety.

The crossing of the Atlantic was a turning point in history, a mark of aviation history and technological sophistication. This clearly was no bunny hop over the dunes of Kitty Hawk. The Atlantic was as ominous an obstacle as any on earth, perhaps the ultimate obstacle. Aviation would soon join trains, ships and automobiles as a serious mode of transportation.

Charles Lindbergh became a worldwide sensation. Similar crowds to the one in Paris would await him in London, Brussels and New York. Charles Lindbergh, the maverick, the unknown, the one who made his mark flying solo, would never be left alone again.

14. Light Bulb
Edison Invents the Light Bulb
October 20, 1879

"It was tedious, backbreaking and heartbreaking," Edison said of his light bulb invention. He literally tried everything from plucking out beard hairs from his annoyed coworkers to threads from spider webs to find a filament that would glow but not burn out.

"Keep trying" he encouraged them. It would take over 1000 trials, each trial taking about a day to hand blow the glass sphere and vacuum out the air. When one of his exasperated workers complained about the many failures, Edison corrected him "we at least know hundreds of things that wouldn't work." Perseverance was one of Edison's calling cards. He once said "Genius is one percent inspiration and ninety-nine percent perspiration."

Thomas Edison (1847-1931)

Edison knew that all things incandense, or glow, but the trick is not to have the glowing filament burn out instantaneously. To this end, he first encased the filament in a globe without air. Without air, and thus oxygen, the filament could not burn, just glow.

October 20, 1879

Finally, On October 20, 1879, Edison and his Menlo Park associates' hard work paid off. They gathered around a small wooden table in his lab much as they had done a thousand times before. He carefully hung the fragile bulb from a wooden beam. "A horseshoe shaped filament of cotton sewing thread impregnated with carbon and sealed in a vacuum" he noted for his note takers. He sat down in his broad wooden chair and applied the current.

"Here goes," he said cautiously as beautiful golden light filled the room. It reflected off Edison's wide forehead and pug nose. "One thousand one, one thousand two, one thousand three," he counted. Co-worker Francis Jell pulled out his customary pocket watch as the seconds started to click. Soon a minute passed and then two. One hour turned to several and a spirit of joy filled the room. Co-workers began singing as Edison told one story after another, all the while keeping one vigilant eye on their new creation. Finally, around noon the next day Edison declared "If it will burn forty hours, it will burn 400" he said. He cranked up the current to find the bulb's upper limit and then sent it to the microscopes. It was a miraculous day.

Ever the showman Edison would wait a bit to reveal his miracle. On December 20, 1879, the media was invited to a night time unveiling of his electric bulb. As the trains pulled into Menlo Park, they were amazed to see a whole little village lit by his new bulbs. As they left one lit building into the next the street lights illuminated their way casting crisp shadows on the snow banks below. One New York Herald reporter described it as "a tiny beautiful light, like the mellow sunset of an Italian autumn."

Edison would go on to create a total of 1098 inventions but this was arguably his greatest. A cleaner, safer, cheaper light source had been found. Edison's success lied in perseverance. As a young boy growing up in Port Huron, Michigan he was full of mischief, and constantly sent home from school. One of his teachers described him as "downright incorrigible." Thank God for that, because without it, the electric light bulb may never have seen the light of day.

15. Television
Philo T. Farnsworth Invents Television
September 7, 1927

It was science project day in Justin Tolman's class in 1920. In came the obligatory solar system model, the baking soda volcano that came to life when vinegar was added, and the endless parade of poster boards. Mr. Tolman looked out the window to the rural countryside of Rigby, Idaho and wished he had chosen to be a farmer. He started to nod off a few times until it was fourteen year old Philo T. Farnsworth's turn. Farnsworth proceeded to write over several blackboards his idea of sending pictures across electromagnetic waves, what would later be known as television.

**Philo T. Farnsworth
(1908-1976)**

In 1901 Marconi shocked the world by sending a radio signal across the Atlantic from Cornwall, England to Newfoundland, Canada. In 1908 Lee Deforest pushed the technology further by sending a human voice over these electromagnetic waves. Who would be the first to send pictures over these carrier waves? Farnsworth thought he had the answer.

Farnsworth got his inspiration while tilling a potato field at his boyhood farm. He reasoned that an image could be translated into sendable data by going row by row with a beam of electrons in less than 1 tenth of a second, much like the way he was tilling the field, or how one mows the lawn. He would read everything he could from his remote farm outside Rigby, Idaho, sometimes by candlelight.

Justin Tolman sat with amazement as young Philo Farnsworth discoursed at length on the theory behind his concept. It was clear to Tolman that Philo had done some serious work on the matter, and to his credit, he forwarded Farnsworth to Brigham Young University (BYU). They were equally impressed but did not know where to go from there. Farnsworth's work was too advanced for even the professors at BYU and so he was forced to figure it out on his own. Unfortunately, his father died the next year ending Farnsworth's formal education.

Seven years passed and Philo kept working on his idea. Finally, when he was just twenty-one years old, on Sept. 7, 1927, Philo scratched a straight line in the center of a completely blackened piece of glass. He shined a powerful hot light behind it creating a very bright line. He asked his brother-in-law, Cliff Gardner, to hold the blackened glass while he went into the other room to watch the receiver. He then asked Cliff to turn the plate ninety degrees to see if the line would do likewise. Cliff did so as Farnsworth instructed and the line moved from horizontal to vertical.

"There you have it" Philo stated matter of factly, "electronic television." His understatement could not hide the sparkle in his eye. George Everson, one of his nervous backers, was a little bit more demonstrative, "well I'll be darned, the damn thing works!'

I'd like to finish this story with a rags to riches ending but such was not the case. There were others working on television and they weren't ready crown Philo king of the hill without a fight. Vladimir Zworykin, a Russian scientist working for RCA, was also close to developing electronic television and filed for a patent in 1923, four years earlier than Farnsworth. A nasty court fight developed. In the U.S. Patent Courts, RCA could show no evidence that they had made any successful transmission in 1923.

In the end, it was Philo Farnsworth's old chemistry teacher, Justin Tolman, that provided the Coup de Gras. He testified on Farnsworth's behalf, that Farnsworth had revealed his ideas in his chemistry class, placing his invention in 1920. This predated Zworkin by three years. Tolman also produced Farnsworth's original science project sketch shown in his classroom many years ago. The court ruled in Farnsworth's favor. RCA's lengthy court appeals and a WWII ban of television sales financially ruined Farnsworth. In addition, his television patents soon expired. Like a waiting vulture, RCA swooped in and mass produced television sets as soon as Farnsworth's patents ended.

Farnsworth received little compensation for his invention and retired to a secluded existence in Maine. Towards the end of his life Farnsworth refused to even let the word television be uttered in his house. There was one exception however. On July 20, 1969, Farnsworth watched with a billion other people as astronauts Neil Armstrong crunched the first human footprint on the surface of the moon. "This makes it all worth while," said Farnsworth. Thanks to Philo T. Farnsworth, the whole world became a global village. His special vision helped to channel the world into every living room.

16. Alternate Genius
Tesla Engineers Alternating Current (AC)
May 1, 1893

When Edison invented his light bulb in 1879 there was no infrastructure of electricity to power these bulbs. He first wired up his laboratory at Menlo Park, NJ for a demonstration and then began to power up New York. By 1882 Pearl St. in downtown New York had electrical power. Edison's power station ran by direct current (DC). In a simple circuit, this meant that electrons start at one terminal of the battery, run through the circuit perhaps lighting a bulb and then finished the one way trip at the other end of the battery.

But Edison's DC power had a fatal flaw. It could not travel long distances. After about a mile or so, the electricity would be exhausted. One would need a power station as frequent and close as fire houses or post offices.

Fortunately, there was an alternative, alternating current (AC). Instead of electrons moving from point A to point B in a circuit, they would alternate directions 60 times a second giving off energy. It was conceptually much more complex but it could travel hundreds of miles without a dissipation of power. AC needed a genius to flush this theory out to practical working motors and generators. This genius was Nikolai Tesla.

Born in Croatia, Tesla was an electrical visionary. He arrived in New York at the age of 26 in 1884. He was hired by Edison to improve the efficiency of Edison's DC generators. Edison's supposedly promised him a bonus of $50,000 if he could make the generators 20% more efficient. Tesla worked feverishly for two years and did just that. He even exceeded the 20%, but then Edison reneged on his promise.

Edison claimed that he was only joking about that bonus and that Tesla did not understand the American sense of humor. Tesla understood just fine but he wasn't laughing. Twenty-nine year old Tesla left and the two were bitter enemies for the rest of their lives. Besides, Tesla had dreams of building AC generators, not Edison's DC ones.

For the next year Tesla dug ditches in New York until he finally scraped together enough money to open up his own laboratory on Liberty Street. Here he worked out most of the fundamentals for AC power including AC motors, generators, transmission lines, and transformers. An astute George Westinghouse immediately bought out his patents and with them started Westinghouse.

In 1888, Westinghouse and Tesla tried to sell the superiority of the AC system. But General Electric, and its main shareholder, Thomas Edison tried to convince the public otherwise. Edison, through surrogates tried to show the dangers of AC current. To demonstrate this, a circus Elephant named Topsy, was publicly electrocuted. They even invented an electric chair powered by AC current to reinforce the dangers of this form of power.

Still Tesla and Westinghouse persisted and with great difficulty and succeeded in securing the contract to light the 1896 World's Fair in Chicago. On May 1, 1893 President Grover Cleveland opened the Chicago World's Fair with a flick of the switch and Tesla's AC power system performed magnificently. Anyone who visited the fair could see that the AC system worked exceedingly well. The doubters were gone and Westinghouse and Tesla were given a contract to harness Niagara Falls. By 1900 Tesla's enormous Niagara generators were supplying power to New York City some

360 miles away. Edison was finished. He had backed the wrong "horsepower" in this race.

In a perfect world Tesla and Edison would have combined their genius as sort of a "dynamo-duo" and produced a technological world of which we can only dream. But this was not a perfect world and the only thing bigger than their collective brain-power were their collective egos. Like AC and DC current, they were quite incompatible with one another. In their power struggle Tesla and Edison helped to bring the twentieth century to light, but as for getting along with each other they unfortunately were very much in the dark.

17. "Saturn Has Ears"
Galileo Introduces the Telescope - 1609

"Be careful of your step," Galileo cautioned as he escorted Antonio Prioli, the Doge of Venice, up the staircase to the top of St. Mark's Tower. It was the highest point in Venice overlooking its fleet of sailing vessels. The year was 1609. Behind him was an endless parade of important Venetians anxious to see Galileo's telescope.

"I think you will like what you see," he said to the Doge as he focused his telescope. It was a twenty-four inch narrow tube that could magnify up to eight times what it was viewing. From the top of the tower one could see all of Venice in her naval glory, magnificent ships and an open bay which was her mercantile jugular.

When the Doge peered into the tube, he was astounded. He could see ships approaching for over two hours before one could see them with the naked eye.

"Why this is staggering!" remarked the Most Serene Signory, another important nobleman. "Two hours to board ships, provision them, and make it out to sea. It makes all the difference. It's no small task moving those vessels from port, you know."

A grateful Venetian senate immediately doubled Galileo's salary and made his appointment to the University of Padua a lifetime position. It made little difference that the original inventor of the telescope was Johann Lippershey from Denmark. It was Galileo that had brought it to Venice and besides, Galileo had made significant improvements.

It wasn't long before Galileo turned his spyglass to the stars. From his little courtyard in Padua, beginning in 1609, he would spend countless nights studying the heavens. When he first saw the rings of Saturn he remarked "Saturn has ears" or bumps on each side. Much of what Galileo saw did not fit with Ptolemy's earth centered universe.

In particular he could see that the moon was not a pristine round sphere as Ptolemy had suggested. It was filled with craters, ridges and mountains. It was a world onto itself. Much of what Galileo saw was supportive of the Copernican sun-centered universe and put Galileo on a collision path with the Catholic Church and history. What the microscope did for biology, the telescope did for astronomy.

Today's telescopes continue to shed new light on our ideas. They have helped to confirm the Big Bang theory, black holes and quasars. The Hubble telescope is so powerful it can read the date off a dime from its perch 373 miles in the air. The telescope would bring into focus the reality of the solar system, and through its light help to correct our myopic vision of the world.

18. Jenner
Dr. Edward Jenner Pioneers the World's First Vaccine
May 14, 1796

When English doctor Edward Jenner left London in 1777 for his rural hometown of Berkeley many considered his promising young career to be minimized. He would be out of the cutting-edge loop they reasoned. As it turns out, fortune would have Dr. Jenner pioneer the world's first vaccine. On May 13, 1796 he inserted cowpox into the arm of eight year old James Phillips in hopes of protecting him from full blown smallpox.

Edward Jenner (1749-1823)

Dr. Edward Jenner dismounted his horse and tied it to the Phillip's rickety gate. He rapped softly on the door and he was soon greeted by an exuberant Mrs. Phillips who escorted him into the parlor.

"The good doctor, won't you come in?" She asked.

"Thank you ma'am I will" he replied. His pleasantries were masking the hideous request he was about to make of his patients. Dr. Jenner got right to the point.

"I want to inoculate your son James with a new vaccine for smallpox." James was their eight-year-old son.

He explained that it was common wisdom that milkmaids that had contracted cowpox while performing their duties, seemed to be immune from the scourge of the much more serious smallpox. By injecting this milder form of the cowpox disease he hoped the natural antibodies would ward off this terrible plague, but he needed to prove it.

It had to be someone who had not had smallpox or who had been preventively treated for it through variolation. This eliminated Dr. Jenner, his immediate family or the other Phillips family members. The current practice of variolation was very dangerous. It gives the patient a mild form of the active virus. Unfortunately, far too many people contracted full blown small pox from this method and died or were scarred for life from it.

Mrs. Phillips did not seem terribly upset. But then Dr. Jenner left out the most horrific part. In order to prove that the body had created successful antibodies from the cowpox, he would need to intentionally inject the live smallpox virus into James at a later date.

"No!" Mrs. Phillips screamed. A distracted James looked up from the floor were he had been playing. "I won't have it!" she sobbed.

Gardener Phillips hurried to his wife's side touching her elbow to support her. She was nearly collapsing. They retreated to the next room. Somewhere in their emotional struggle the Phillips had a courageous moment and agreed. Mrs. Phillips own mother and sister had died during variolation and she dreaded it almost as

In 1796 Sarah Nelmes, a local milkmaid, had contracted cowpox from her cow Blossom

much as Jenner's proposal.

Dr. Jenner lost no time. The next day an unwilling James was brought to the Chantry Cottage where Dr. Jenner lived. He did not fully comprehend the danger he was in but he was quite sure that it involved a needle and that was enough for him.

Sarah Nelmes, a local milkmaid, had contracted cowpox from her cow, Blossom, and was suffering from its ill effects. Dr. Jenner lanced a particularly bad pustule on Sarah's wrist and removed its gruesome contents. He turned toward James and skillfully scratched the child's arm enough to deposit the virus beneath the skin. A wobbly Mrs. Phillips needed more restraint than James.

As expected several days later James developed the anticipated cowpox and became ill but he soon recovered. In July, however, came the moment of truth. Dr. Jenner loaded up a lancet with a potent dose of the smallpox disease. He looked more like he was holding a loaded gun, than a small needle, and he was. Without delay he applied the deadly virus to the young boy's arm. It was done.

Eight days later, the normal incubation period for small pox, James showed no signs of the smallpox disease. Eight days became ten, and soon two weeks had passed without so much as an elevated fever. Dr. Jenner began to hope. Each day that passed found Mrs. Phillips in a happier mood than the day before. Soon it became clear, the miracle had worked. Jenner would call it a "vaccination" after the Latin word "vacca" which means "cow."

The news of Jenner's miracle vaccine spread around the world. Some newspapers lampooned him, drawing cartoons of people turning into cows after one of Jenner's cowpox vaccinations. From 1800 to 1900 the number of smallpox cases dropped from forty out of every ten thousand people to one in every ten thousand. Others vaccines would follow; rabies, diphtheria, tetanus, pertussis, polio, meningitis, measles and rubella to name just a few. Jenner's discovery of vaccines on 1796 is was one of the most important moments in science and medicine in the history of mankind.

Some newspapers drew cartoons of people turning into cows after one of Jenner's cowpox inoculations.

19. Mendeleev
Mendeleev Invents the Periodic Table
1869

The science of chemistry in 1869 was still in its adolescence. Chemists were fumbling to find a pattern in the chaos of seemingly unrelated elements which in 1869 numbered only sixty. Today we have over 117 elements. Thirty five year old chemistry teacher, Dimitri Mendeleev would bring order to the elements on a train bound for St. Petersburg. He would invent the Periodic Table.

The St. Petersburg express whistled twice as it pulled out of the Moscow station. Russian Chemist Demetri Mendeleev settled into his 3rd class seat balancing his customary cup of tea and a stack of index cards. "It would be a nice ride," he thought, a chance to see the countryside that he loved so much.

He could have afforded a first class ticket but he preferred coach. He enjoyed talking to the peasant farmers about the latest advances in agriculture and perhaps he felt a kinship with their hardships. He had grown up in Siberia, the last of fourteen children. His mother and father were both deceased.

Mendeleev laid out the index cards on the table in a kind of solitaire he had been playing unsuccessfully for nine years. He ruffled his long white beard as he vowed to break nature's hidden structure. "It's there, I know its there," he mumbled to himself. He saw that every seven or eight elements the properties would go from metallic to non-metallic and then repeat. There was a repeating pattern to the elements.

Dimitri Mendeleev
1834-1907

Finally, it hit him. The secret is in the atomic weights. He organized the elements from lowest atomic weight to the highest. Each time a series of elements began to repeat its metallic properties he would make a new row. Those elements that showed similar properties were called "families." He published his table of elements on March 6, 1869.

Unlike others working on a similar theme, Mendeleev left gaps in his table for yet undiscovered elements. Three of these gaps were for gallium, scandium and germanium. These elements were soon found and exhibited exactly the properties Mendeleev had predicted by his table. He also pointed out several elements that he believe were in error according to his table and further testing confirmed this. Mendeleev's Periodic Table became the standard in modern chemistry. In 1913 Henry Moseley adjusted Mendeleev's original insight to use atomic number instead of atomic weight but the original order of the table into its periods and families remained intact.

Today every high school chemistry student has a copy of Mendeleev's periodic table if not in front of him, certainly on the wall close by. It is the starting point of anyone trying to understand the chemical world. Mendeleev was caught up in the political unrest of Russia and therefore lost his job in 1870. It is comforting to know that at least in his small corner of the world, chemistry, he found stability, order, rhyme and reason to a political world that sorely lacked it.

20. Goddard
Goddard Pioneers Modern Rocketry
March 16, 1926

"Looks great" thought 44-year old Robert Goddard, as he backed a safe distance away from his homemade rocket.

His Auburn, Massachusetts neighbors grumbled as they pulled back their curtains in this thickly settled neighborhood.

"It is four months until the fourth of July, a bit early for fireworks don't you think?"

"Fireworks," Goddard would be insulted to hear his liquid rocket labeled as such a crude device. This was no Roman candle, once lit, a firecracker burns to completion. One can't light half a firecracker. But this, "Nelly" as he calls it, is a liquid rocket. By deciding exactly how much propellant gets used when and where affords the engineer control over its flight. It was the first of its kind and

Dr. Robert Goddard, a professor at Clark University, was the world's leading authority.

It was March 16, 1926 on cold snow-covered day at his Aunt Effie's farm just outside of Worcester, Massachusetts. Professor Goddard (Oct. 5, 1882-Aug.9, 1945) looked over the eight-foot high tripod holding his rocket and then lit the fuse.

His right hand woman was his devoted wife Esther who aimed her home movie camera to capture the rocket's historic path. "Four, three, two, one, shhhhhhhh!" His rocket rushed upwards, sending a plume of white smoke into the cold air before landing in a cabbage patch.

"Two point five seconds," an exhilarated Goddard announced. "The distance is 184 feet." He would later meticulously record his data in his very detailed notebooks.

Robert H. Goddard (1882- 1945)

His rockets would grow over the next several months as would the ire of his neighbors who wished to send more than his rockets to the moon. "Crack-pot," some labeled him. Others called him a "hopeless dreamer." The Worcester Telegram headlined "Moon Rocket Misses Target by 238,799 ½ miles" poking fun at the difference between the height Goddard's rocket had achieved, half a mile, and the remaining distance to the moon. The New York Times jumped on the band wagon goading Goddard saying, "Everyone knows rockets won't travel in space because there is nothing to push against."

On one particularly bad day on July 17, 1929 one of his rockets blew-up causing a fire and great distress in the neighborhood. It was covered heavily by the local news media. The windows of the surrounding houses were rattled. One woman thought a plane had crashed in her back yard.

"Enough is enough" cried neighbors who said, "He has to go." Goddard agreed. He too was ready to go. He needed to find a place where there weren't so many houses. He needed some open space, a place where he was free to crash and burn, but also a place he could learn and grow. That place turned out to be Roswell, New Mexico. It was a place that had more cows than people. The warm dry climate would also be more forgiving of his poor health as Goddard battled a lifetime disease of tuberculosis.

But this kind of move to Roswell and research would require money, more money than his professorship at Clark University would allow. As it turned out the fire of 1929 was a stroke of luck. While it assured him of a short stay in Auburn, Massachusetts it had made the national news and caught the attention of the famous pilot Charles Lindbergh. Being the first to fly solo across the Atlantic, he knew something about long odds, determination and being alone. In Goddard he found a kindred spirit. With Lindbergh's help Goddard secured a $100,000 grant through the Guggenheim foundation. In 1930, Goddard, his wife, and four helpers packed their rockets and made the move to Roswell, New Mexico.

In New Mexico, Goddard's rockets grew by leaps and bounds and by 1931 his rockets reached heights of a mile and a half and speeds of 500 mph. By 1935 this altitude would triple to over 7,000 feet. His three-foot rockets were now more than eighteen feet and he developed many of the control mechanisms that were imperative to any successful launch. He successfully filed for dozens of patents over nine years.

Despite Goddard's attempts to interest the United States Government in his work, they were simply not interested. He did everything possible to call their attention to what he considered a National emergency but no one cared, at least in this country. On the other side of the Atlantic, the Germans were interested, very interested. By the mid-1930's Germany had built a menacing war machine and armed conflict was inevitable. Germany had fallen under the spell of the great orator Adolph Hitler and his Nazi regime and they sought Goddard's help to rain "thousands and thousands of bombs on England," to quote Hitler. Goddard, being a loyal American citizen and patriot, promptly refused. The Nazi's got all the help they needed from the United States Patent office which simply sold copies of Goddard's patents for a mere thirty-five cents apiece. It was unthinkable and it occurred without Goddard's knowledge.

Under the watchful eye of young genius Werner Von Braun the Nazis used Goddard's patents to build their infamous V-2 rockets which killed thousands in London and brought terror on the streets. It nearly won the war for them. Fortunately, the development of the V-2 rocket was very late in the tide of events of WWII and the rocket sites were soon overrun by General Patton's streaking third army. The Germans also had plans called Project America that would have sent similar V-2 rockets to Washington, New York and Boston. It was perhaps only a year a way.

Goddard died on Aug 9, 1945 at the age of 63 from cancer of the larynx. Unappreciated by his country, his devoted wife fought for the recognition that Robert truly deserved. She was awarded a one million-dollar settlement in 1960 from the United States government for rights to more than 200 patents covering "basic inventions in the field of rockets, guided missiles and space exploration."

Goddard never expected his own work to reach the moon. He knew that such an enterprise would take a committed government with vast resources several generations. This would happen two decades later under President Kennedy and the Apollo program.

A few days after Neil Armstrong and Buzz Aldrin walked on the moon, the New York Times printed an apology to Goddard for its past derision. Aunt Effie's farm in Auburn, Massachusetts is now a golf course. A whole different variety of small projectiles, golf balls, make their way through the open skies overlooking Worcester. Every now and then an errant sportsman shanks a nine-iron and lands behind a waist high monument in the shape of a rocket. Its inscription reads, "Here Robert Goddard launched the first successful liquid rocket on March 16, 1926." Goddard, like his monument, was quiet, steadfast, unassuming, and eternally pointing upwards.

21. Dolly
Scientists Clone the First Mammal - July 5, 1996

On February 22, 1997, Dolly danced sheepishly on her lamb hooves trying to make sense of the popping flash bulbs and reeling film. She was big news. Dolly was the world's first cloned mammal. She was the product of the Roslin Institute of Scotland, directed by Ian Wilmut.

Dolly was a white-faced Finn Dorsett lamb born on July 5, 1996. She is a white-faced lamb born to a black-faced mother. The implication of this fact guarantees that Dolly was not conceived by conventional mating. Since black-faced genes are dominant, it would be impossible for its black-faced mother to have given birth to a white-faced lamb. A DNA analysis of Dolly later confirmed that she is genetically identical with her donor, not her mother.

Cells from Dolly's mother were taken from her udder. The nucleus of these cells was removed and replaced with genetic material from Dolly's donor. It is significant because it was thought that cloning could not take place with adult cells. They had to reinvigorate the adult cell by shocking it with electricity. Wilmut's ordinary tone stood in sharp contrast to the extraordinary efforts needed to produce Dolly. For the one success of Dolly there were 277 failed embryos!

Dolly lived six years before succumbing to a common lung disease in 2003. Her life expectancy was a bit shorter than the average but she also had a great deal of contact with the outside world. Visitors of every kind came to see this medical miracle. She gave birth to six lambs in her lifetime. After her death she was stuffed and mounted. Today her remains are on display at the Royal Museum of Scotland.

Was this a great moment of science? The ethical questions and value of cloning are enormous. Cloning genetically identical mice for medical research is invaluable. In fact several mice have even been patented. One such example is the "oncomouse" patented in 1998, a mouse especially cloned for cancer study, or "oncology."

Using cloning, it might one day be possible to bring back an extinct species as in the 1990's blockbuster movie Jurassic Park. Scientists were able to clone extinct dinosaurs from DNA found in a well preserved mosquito. Most feel it is doubtful that any DNA can last that long so it is improbable.

One could envision a grief-stricken pet owner cloning their lost pet, or perhaps a lost child. In the 1978 movie "Boys from Brazil" someone tries to clone a whole race of Adolph Hitlers. In planning a family it could be tempting to want another Einstein, Mozart or a Ted Williams.

Since Dolly, other scientists have cloned cats, mice, cows, goats and pigs. Several are trying to clone horses and dogs. In the movie Jurassic Park Dr. Malcolm raises the bioethical question around cloning when he says, "Scientists were so preoccupied with whether or not they could; they never stopped to think whether or not they should."

22. The Transistor
Bardeen, Brattain and Shockley Invent the Transistor
December 16, 1947

What exactly is a transistor? It's a small three pronged electrical part that fits in the palm of your hand. It is used to amplify current. It replaced the bulky vacuum tubes that performed the same function but were much larger and subject to breakage. The vacuum tubes looked like small refrigerator light bulbs only more cylindrical. Like all bulbs they were subject to the same failure rate. Vacuum tubes and light bulbs were vulnerable to dropping and vibration. Also, the filaments would burn out at the worst possible times. The average television set of the 1950's had maybe 100 or so of them. Because of their unreliability and short life-span, the search was on for a replacement to these tubes.

Their replacement was called the "transistor," a small part about the size of the top button of a retractable pen. It had no moving parts to jiggle, no filament to burn out and no glass. It was virtually indestructible. They would last as long as the rest of the device if not longer. Its invention ushered in an unprecedented time of miniaturization. It was the age of solid state.

Ground zero for inventing this new device was Bell Labs in New Jersey. Bell Labs appointed William Shockley to head up a team that included John Bardeen and Walter Brattain. Science loves an underdog, and it was actually John Bardeen and Walter Brattain, Shockley's subordinates who would invent the transistor, not Shockley.

John Bardeen was educated at Princeton and was the theoretical soul of the team. He had skipped three grades and entered college at the age of 15. He was nicknamed "whispering John" because of his quiet demeanor and even temperament.

Walter Brattain was the mechanical genius. They say he could fix or build just about anything. He was helpful and had a kind grandfather like quality which made him easy to work with.

One night on December 16, 1947, Brattain and Bardeen were working on the surface resistance problem of the new device when Walter Brattain noticed the graph on the oscilloscope.

"This has gain!" he muttered to himself, electronic code words for it has amplification. This is exactly what they had been looking for. Working late into the night Brattain and Bardeen came up with a working transistor.

When Bardeen and Brattain called Shockley at his home to tell him of the great news, Shockley was stunned. He was outwardly happy of course but he felt that he was destined to invent the transistor, not his assistants. Shockley realized that although Bardeen and Brattain had made the historical invention it was a weak design. Secretly, Shockley sought to improve on their original invention and within four weeks he succeeded. It was a much more stable design and would be the used by the world. On June 30, 1948 a Bell Labs spokesman made public their invention.

Bardeen and Brattain were hurt that Shockley had not included them in the development of the new design and upset when Shockley tried to file for the patent as the sole inventor. They became bitter enemies and both Shockley and Bardeen left Bell Labs shortly afterwards. Bardeen would later receive a Nobel Prize for his work on superconductivity.

Anything built after 1950 that was smaller, faster and better probably owed much of its improvement to the transistor. It was in many modern day products. It made feasible small transistor radios that could work on batteries for a long period of time. Several billion dollar companies such as Sony and Intel got their start with the transistor. We all know that silicon chips are the basis for our computers, but it should be noted that on each of these postage sized parts are millions of transistors. From cell phones to handheld calculators to iPods; all of these would not exist with the transistor.

In 1956, the Nobel Prize in physics was awarded to John Bardeen, Walter Brattain, and William Shockley for their invention of the transistor. During the ceremony the threesome was praised for their teamwork and close cooperation. Although this was not the case, at least both Bardeen and Brattain were recognized for their contribution to one of the most important discoveries of the twentieth century.

23. Telegraph
Samuel Morse Invents Electronic Messaging
May 24, 1844

Before Samuel Morse the speed of a message traveled as fast as the horses carrying it on the Pony Express. In its heydays of the 1860's the Express could deliver mail via horseback from Saint Joseph, Missouri to San Francisco in just under 10 days, a trip that would normally take 6 weeks. It was an extraordinary achievement. But soon Samuel Morse's telegraph would make it obsolete.

Electronic signals traveling at the speed of electricity would prove too much for both man and muscle.

On May 24, 1844, Samuel Morse sat down at a small desk in Baltimore to send the first electronic message to Washington D.C. a distance of 38 miles. Each signal would take less than a second. Morse had developed his own code, Morse code, which would transmit letters by either short or long buzzes known as dots and dashes. A small crowd gathered round an anxious Morse as he began to tap out the buzzing key. Annie Ellsworth, the young daughter of his friend and backer at the patent office, had suggested a quote from the bible, Book of Numbers 23:23.

"What hath God wrought!" he tapped. At the Capital in Washington DC, his partner Alfred Vail reiterated the same message back and the age of instant communication was born. No other invention, with the exception of the telephone, conquered time and space more than the telegraph. Within ten years every small town in America had a station and tens of thousands of miles of wire had been nailed to trees across the county. By 1869 Morse's telegraph wires stretched all the way to San Francisco. During the Civil War, generals would make use of the telegraph to deploy troops or to inform a beleaguered Abraham Lincoln of the progress of the Civil War.

As with any new technology some people were slow to appreciate its' significance. One of these was murderer John Tawell in England in 1845. After poisoning Sarah Hart, he quickly hopped on a train to make his get away. To his shock he found several authorities waiting for him at the next station. An alert railroad employee had telegraphed ahead his arrival, his description, and even his seat number. As John Tawell found out the hard way, what may look like the "cutting edge," may only be one side of a two edged sword.

24. Telephone
Bell Invents the Telephone - March 10, 1876

The race for the talking telegraph, or telephone, began the moment Samuel Morse invented the telegraph in 1844. The world was impatient to send the spoken word through the wires. It was ironically a teacher of the deaf, Alexander Graham Bell, who would provide this technology.

Bell knew he was close in the spring of 1885, while working in his Boston laboratory with his nuts and bolts guy, Thomas Watson. Watson accidentally twanged a spring and Bell heard it two floors down. At that moment he knew then that it was possible to send the human voice. If the current could recreate the variations of sound made by the twanging of a spring it could recreate a southern twang as well. Within twenty-four hours Watson and Bell had a basic design that worked, not perfectly, but it worked.

Alexander Graham Bell (1847-1922)

"Patent it!" demanded nervous investor Gardner Hubbard. He was also Bell's future stepfather. Bell insisted that it was not ready.

Hubbard knew that Midwest inventor Elijah Gray and others were in hot pursuit on this same invention. Gray had already beaten Bell to the buzzer, so to speak, over the harmonic telegraph and Hubbard did not want to see it happen again. A very nervous Gardner Hubbard filed a patent for the telephone on behalf of Bell on March 10, 1876 just two hours before Elijah Gray.

Bell received his patent #174465, on March 7, 1876. It is considered one of the most valuable patents ever issued by that office. The only problem was that the Bell phone didn't really work. It could transmit the human voice but it was mumbled and barely audible.

The breakthrough came just three days later on the evening of March 10, 1876. Bell looked out of his attic window at 5 Exeter Place, in Boston. He removed his suit jacket and gingerly placed it on the back of the chair and sat down. His new telephone looked like an ordinary black cup with just a couple of wires coming out the back. Two floors down, Watson held an identical device awaiting Bell's call. According to Watson, Bell leaned forward to speak into the cup and his left arm knocked over a jar of acidic water onto his legs. As the acid started to burn, Bell leaped up and screamed

"Watson, Come here! I need you!"

Watson came to Bell's aid. After he was assured that Bell was not seriously injured, they both realized the phone had worked.

Within a year Gardner and Bell formed the Bell Telephone Company and over 600 phones were installed in Boston. Busy bankers were anxious to stay in contact with their offices.

In 1914, Watson and Bell hooked up for the first coast to coast call between New York and San Francisco. Bell reiterated his famous first call "Watson, come here!" telephoned Bell.

"I would be happy to Mr. Bell," Watson replied, "But it will take me over a week to get there."

Thanks to a quick thinking Gardner Hubbard, fame and fortune was secured for Bell and Elijah Gray became just a footnote in history. Bell's invention is one of the greatest in human history. Thomas Edison said of the telephone, "It annihilated time and space and brought the human family closer together."

25. Marconi Miracle
Marconi Sends a Radio Signal across the Atlantic Dec. 12, 1901

On Dec. 12, 1901, the wind whistled and pinged through the old Newfoundland hospital unimpeded by the stucco walls. Twenty-seven years old Giglielmo Marconi (1874-1937) hunched his narrow shoulders but would not be distracted from the task at hand; to send the first radio signal across the Atlantic Ocean. The current world record for radio signals was eighty miles. Marconi's feat, should he succeed, would be over 2,000 miles from St. Johns, Newfoundland to Cornwall, England.

"Newfoundland is a God awful place" he thought "Not a bit like my native Italy." It did have one geographical plus, however, it was the

Gigliemo Marconi (1874-1937)

most easterly North American point to Cornwall, England which more than made up for its lack of ambience.

Marconi and his team nervously checked and rechecked his receiver in anticipation of this historic first transmission planned for 12:00 noon Newfoundland time, which is 3:00 p.m. Cornwall, England time.

Marconi opened and closed his hands trying to keep them warm. Atop Signal hill he can see the vast and foreboding Atlantic Ocean stretching for as long as the eye can see. His two long time assistants, George S. Kemp and Percy Wright Paget, looked equally cold but were dressed more suitably than the dapper Marconi who customarily wore tailored three piece suits.

"Impossible," Edison had dubbed the project. Many experts shared this view. The curvature of the earth was the problem. For every mile one goes out, the earth curves downward about a meter or yardstick. Over 2,000 miles, the distance of this transmission, this adds up to about a mile drop due to the curvature of the earth.

The argument was that this would cause the radio waves to go over the heads of the receiving station and harmlessly into space. Marconi hoped instead that the signals would hug mother earth and curve with it. Marconi would get an unexpected bit of luck from an undiscovered ionosphere about 100 miles above the surface that absorbed and bounced back radio waves to the surface. Marconi's secret plan was to use huge eight foot tall kites to carry an antenna aloft to catch any signals sent from Cornwall. The extra height provided by the kites would help to compensate for the curvature of the earth problem.

Earlier this morning, the ferocious Newfoundland winds sucked the kites skyward and they were quickly lost in the clouds above. The team had agreed upon a simple message, the letter "S" to be sent in Morse code, which is three short sounds or dot-dot-dot as typed on a conventional telegraph.

Just prior to 12:00 Marconi sat down for what he hoped would be a momentous occurrence. He pressed the receiver to his ear and put his finger in the other ear hoping to capture the faint bleeps. His crude radio receiver was no bigger than a breadbox, mostly of wood with brass knobs. It looked more like an ornate music box.

At 12:31 in the afternoon, Marconi leaned forward a little as if that would help his hearing.

"I can hear it" he said, "I can hear it!" Softly, nearly imperceptible, but he had heard it. He quickly handed the earpiece to Kemp. "Do you hear it?" Marconi asked. Kemp listened.

"Yes," Kemp said, "It is the letter S, but it's very weak.

They repeated the experiment the next day and once again heard a very faint "S." Marconi informed the press the next day. "A Miracle!" one newspaper headlined. A red faced Thomas Edison tried to save face by commenting that he was glad the young man had done it because it saved him the trouble.

But while the news media hailed his signal as a great technological victory it was a bit premature. All Marconi had for evidence was his word and Kemp's that they had in fact heard the signal. A much more inclusive proof would need to be offered. It would be six more long years before Marconi worked out the technical difficulties so that reliable clearly audible messages could be sent worldwide. He would make the signal wavelength longer, and the antenna shorter and more directional. He would also discover that night transmissions were much easier. During the day the sunlight pushed the ionosphere closer to the earth which absorbed some of the signal. Had he sent his famous Dec. 12, 1901 letter "S" during the night he would have had a much stronger signal.

Cabot Tower

On Oct. 17, 1907 Marconi finally believed his radio system could do what he said it could, that is reliably transmit radio signals across the Atlantic. He began transmitting and receiving radio signals between Cliften, Ireland and Glace Bay, Canada. Over 10,000 words were sent in the first day. At the end, an exhausted but happy Marconi collapsed into his chair. Today the hospital on Signal hill is a parking lot. The original building burned down in 1920 and was not rebuilt. Nearby is Cabot Tower that overlooks the seemingly endless North Atlantic.

Wireless radio proved a godsend to the shipping industry. On July 23, 1909 two ships, the S.S. Republic and the S.S. Florida collided on a foggy night. A panicked telegraph operator radioed Nantucket for help and 1,650 lives were saved. On April 14, 1912, another ship struck an iceberg and was sinking. While 1,013 passengers were tragically lost, 700 people were saved thanks to radio telgraphy. That ship was the Titanic.

26. "And Yet It Moves!"
Galileo Argues for a Sun Centered Universe
1633

On June 22, 1633, a frail 70 year old Galileo Galilei, the most famous scientist in Italy, in fact the world, shuffled into the Inquisition Hall dragging his heavy leg irons. He was here to accept Rome's punishment for preaching heresy. Specifically it was for preaching about a sun-centered universe, a model known as the *heliocentric* model. The Church stated that man was the center of God's creation and therefore earth, not the sun, had to be in the center. God would not have put earth in a secondary position in the solar system circling another sun. The accepted model was that of Greek astronomer Ptolemy from 150 AD. The earth was in the middle of the solar system with the six known planets and the sun orbiting around it. It was known as the *geocentric* model.

"And yet it moves"

A broken Galileo could not believe that it had come to this. He was not the first one to express these views. It was Copernicus that placed the sun in the center, but he did so in a book not published until his death.

It was left to Galileo to argue for the sun centered universe. This he did magnificently in his book *Dialogue Concerning the Two Chief World Systems.* He showed how Venus had phases like the moon. He explained how the moon was not a boring smooth sphere, but pockmarked with numerous craters and mountains. It was another world. He showed how the spots on the sun moved across its surface, which is consistent with a turning orb, not a stationary one. All of this pointed to a sun centered universe.

Galileo had taken great pains not to offend the Catholic Church. His book was written as a hypothetical, a "what if?" It was a spirited dialogue between three friends Simplicio, Salviati and Sagredo. But the advocate for the Church's position was named Simplicio. The resemblance of

Simplicio's name to the word simpleton was not lost on the Church. To add insult to injury, Simplicio, the simpleton, mouthed many of the Pope's own words. An infuriated Pope Urban III was determined to put Galileo and the earth back in their rightful places.

Galileo was noticeably shaken. He was aware of the fate of poor Father Bruno who was burnt at the stake in 1600 for holding such heliocentric views. A penitent Galileo knelt before the ten judges and publicly denounced his sins of preaching the Copernican doctrine. He argued that he never intended the Copernican theory to be his own, that it was just a literary tool. No one in the room believed him, but the lip service was important.

The court announced its sentence. He would be under house arrest for the rest of his life. Galileo returned to Florence where he spent his remaining days at a villa called Arcetri. It was humiliating and restrictive. Ironically, the years spent at Arcetri, despite his house arrest and ill health, were amongst his most important and fruitful. It was here that he did his work with inclined planes and acceleration. His later work, *Dialogue Concerning Two New Sciences*, became the cornerstone of Newtonian physics.

Johannes Kepler (1571-1630)

Pope John Paul II would eventually vindicate Galileo in 1982. "Mistakes were made," he said. As Galileo left the room, legend has it, the scientist in him could no longer be suppressed. As he reached the doorway, he turned, and quietly mumbled "Eppur si muove" ("And yet it moves") referring to the motion of the earth. Despite the papal banning, Galileo's *Dialogue* was quickly smuggled throughout Europe and widely accepted. Soon the Copernican theory won over and the heliocentric universe would have its day in the sun.

27. Kepler
Kepler Makes the Orbits Ellipses 1609

1596 Johannes Kepler thought he had read the mind of God. He believed that God had imbedded perfect shapes into the orbits of the planets.

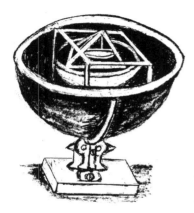

Perfect Shapes Kepler believed God had imbedded geometric shapes into the orbits of the planets

Each planet's orbit had its own particular geometric shape. For example, the orbit of Saturn is actually the equator of an enormous invisible sphere. The orbit of Jupiter is a circle that is inscribed in a square. Other planets like Mars and the Earth have circles imbedded in triangles and tetrahedrons. Together these shapes formed a set of Chinese boxes where each is smaller than the last. They all compactly fit together within the solar system in a "Harmony of the Spheres" or "Harmonice Mundi" as Kepler called it in 1596. Yet Kepler's shapes did not exactly match the known positions of the planets over the year. He felt sure that if more accurate enough data were found his perfect shapes would be in perfect agreement with the data.

Kepler needed better planetary data and he knew exactly where to get it, a hard drinking hard living Danish astronomer Tycho Brahe. He joined on with Tycho at his massive observatory in Prague, but soon was badly disillusioned when Tycho refused to give up the pertinent information. Tycho was combative, short-tempered, and insulting. The worst was that although Tycho had all the data in the world, he wasn't sharing it. He would only occasionally let slip the proper positions of Mars over dinner table just to tantalize Kepler but he would never fully disclose the full volume of his observations.

Finally after 18 months Tycho died in 1601. Kepler inherited his massive amounts of accurate planetary data. Unfortunately, instead of proving his theory of perfect shapes, it did the opposite. The correlation between the position of the planets and those predicted by Kepler's perfect shapes was worse than ever.

In one of the most courageous pivotal changes of heart, Kepler abandoned his perfect shapes in favor of a whole new system. He allowed the data to tell him what the shape of the orbits should be, not the other way around. To his amazement, the perfect geometric shape for planetary orbits turned out to be ellipses. The sun was positioned as one of the foci in all cases.

"Oh horror of horrors" he thought "they can't be an ellipses!" In a hold over from 2000 years ago and Pythagoras, an ellipse was an imperfect shape. It is a circle gone bad. Kepler tried for nine years for the data to say something other than ellipses but it was to no avail. Reluctantly he admitted that planetary orbits are ellipses. In the end his imperfect ellipses would provide the cornerstone to Newton's law of Universal Gravitation and bring "harmony" to our solar system model.

Einstein discovers that Time is money

28. Special Relativity
What Is So Special About Special Relativity?
Time-Dilation - 1905

It was an outrageous claim. "Moving clocks run slow," Einstein claimed in a 1905 paper known as the Special Theory of Relativity. Time "dilated" or passed more slowly for moving observers then observers at rest. It was the overthrow of absolute time. We are not all moving forward into the future at the same rate. If one astronaut is zipped off into space at close to the speed of light he would return much younger than the astronaut who remained on earth.

In 1971, Scientists J.C. Hafele and Richard E. Keating set out to test this bizarre "time-dilation" claim. They started with two atomic clocks. Atomic clocks are accurate to within one billionth of a billionth of a second. One clock flew east about a commercial jetliner, while the other atomic clock stayed at home, that is, at the U.S. Naval Observatory in Washington D.C.

To their amazement, the clock that was moving passed time more slowly than the one left at rest. How much time are we talking about? The amount was very tiny since the velocities involved were tiny compared to the speed of light. It was only a few nanoseconds difference, but exactly as predicted. They repeated the experiment flying west and again the moving clock aged more slowly than the clock at rest. The experiment was repeated in 1996 with similar confirmations.

Another more dramatic demonstration is that of the mu-meson, a particle that lives for 2×10^{-6} seconds. It decays into some other particle after that. Since it is 10,000 meters to the earth's surface from where they are created in the upper atmosphere, they should never reach the surface but they do. They are traveling so fast that time-dilates and they live long enough. For Einstein, "The distinction between past, present and future is only a stubbornly persistent illusion," illusion being the keyword.

29. Tycho Brahe
Tycho Brahe Maps the Night Sky
1570-1601

Tycho Brahe (1546-1601)

In 1543 Copernicus finally put the sun in the middle of the solar system where it is. The planets, including the earth, ran around the sun in circular orbits. We had come a long way since Ptolemy's earth-centered solar system but the model still did not match the observable data of the planets. To compound the problem much of the data of the positions of the planets was simply wrong.

Enter Tycho Brahe, a Dutch nobleman turned astronomer who was obsessed with accuracy. He would spend the next 30 years of his life mapping the night sky on a large 5 foot globe at his island observatory at Hven, in the Baltic Sea. He built many of his own astronomy tools including a 38-foot diameter quadrant, or quarter circle, with which to measure angles, but no telescope. All of Tycho's measurements were done painstakingly with the human eye.

Most find the job of astronomy in the 1600's to be somewhat boring, especially minus the telescope but Tycho's life was anything but. He is perhaps one of the most colorful characters in science history. Tycho was a very large man and he loved to throw huge feasts where he would eat and drink to excess. These lavish spectacles often appalled visitors to the island for their lack of etiquette but no one dared to complain. If one did they would simply be escorted to the complaint rooms downstairs which were complete with bars on the windows and shackles. Tycho had a legendary bad temper. In 1566 he lost his nose in dual with a fellow student, Manderup Parsbjerg, over who was the better mathematician. Since that time he wore a silver nose which he often removed during the course of meals to rub ointment on it. Also in attendance at these meals was his friend, a dwarf named Jepp, who would sit under the table coming out every now and then to provide some mischief by which to entertain Tycho.

A prim and proper Johannes Kepler joined this merry crew in search of Tycho's data. He could scarcely endure it and resigned several times only to be drawn back. Eventually Tycho died and Kelpler would use Tycho's observations as the bedrock for the solar system model. It was Kepler's faith in Tycho's accuracy which convinced Kepler that the orbits of the planets had to be ellipses and not circles. "Let me not to have lived in vain" Tycho lamented at the end of his life. Did he live in vain? Hardly, his work was critical. As for the rest of it, he became both famous and infamous at the same time.

In 1851 Foucault would later use a pendulum hung from the Pantheon to prove the earth moves.

30. Galileo's Pendulum 1583

On one Sunday in 1583, nineteen year old Galileo Galilei sat in the Cathedral of Pisa, Italy. He became bored with the service and began looking around when a swinging lamp caught his eye. Perhaps an errant wind had set it in motion.

The lamp hung from a long chain suspended from the Cathedral's very high ceilings. He watched it rhythmically move back and forth and noted that each swing took the same amount of time. After a few minutes the lamp lost some of its energy and the size of each swing diminished greatly. This was not surprising to Galileo but what was surprising was that the diminished arcs seem to have exactly the same period of oscillation as the larger arcs had minutes before.

"This is curious" he thought. Having no watch the ever resourceful Galileo put his fingers to his wrist and used his pulse as a time keeper. It was true; the period of time between swings did not change despite the diminished size of the swings. Of course Galileo being the great scientist that he was followed up this original insight with exhaustive testing in his laboratory with pendulums of every sort. He had made his first scientific discovery thanks to a few moments of reflection and a little inspiration from above.

31. General Relativity
Einstein's General Theory of Relativity
1916

Just as the world was starting to comprehend the ramifications of Einstein's Special Theory of Relativity of 1905, the once Swiss patent clerk shook the universe a second time by his even more ambitious General Theory of Relativity as outlined in his 1916 paper. Ignoring advice to avoid the problem of gravity, Einstein tackled it head long in this theory. Einstein equated gravity to acceleration in a principle known as "equivalence." There was no way to distinguish one from the other. They are two sides of the same coin.

In addition Einstein reframed the universe in a four-dimensional space-time, not the traditional 3 spatial dimensions and an independent time. Time is given equal status to the special dimensions, left, right, up, down, forward and back. In it, gravity is not a force but a shape in this 4 dimensional space-time continuum. Matter warps or curves space-time like a heavy billiard ball on a thinly stretched piece of fabric. This warping causes planets to fall into the center or orbit around it.

Einstein did not mean for his four dimensional world to be merely a mathematical tool or a construct but believed the universe truly was 4-dimensional. It was Einstein's most ambitious work to date but not his last. He would spend the next fifty years of his life in search of a Unified Field Theory. He would attempt to show that the four forces of nature, gravity, electromagnetism, the strong and weak forces were all different manifestations of the same underlying force.

There was gravity, the longest acting force, but despite its long arms, it was the weakest of the four. The strongest of the four forces was electromagnetism which controls all of chemistry. Finally there are the strong nuclear force which binds the nucleus of the atom together and the weak force which is responsible for radioactivity.

In fairness to Einstein, many of the pieces of the puzzle for a Unified Field Theory did not exist at the time of Einstein. Today we are still in search for this "theory of everything" as it has been renamed. It is the Holy Grail of physics, not yet secured.

32. Trieste
The Trieste Journeys to the Bottom of the Sea
Jan 23, 1960

There is a place considered far more formidable and dangerous than the moon. It is not so far away as our distant satellite, in fact it is as close as our backyard. It is the oceans. Not all oceans are the same depth. If one is going to have to walk the plank due to some blood thirsty pirates, the Atlantic is much shallower than the Pacific. The Atlantic Ocean is about three and a half miles at its deepest while the Pacific Ocean can reach depths of up to seven and one half miles at a place called the Marianas Trench. On Jan 23, 1960 two brave Argonauts, Jacques Piccard and Lt. Donald Walsh were the first humans to try to reach these incredible depths in their cigar shaped craft known as the Trieste.

Jacques Piccard and Lt. Don Walsh stepped into a tiny 5-foot wide metal sphere mounted below a 17-foot cigar shaped float. Jacques father, Auguste Picard, the famed balloonist, had designed it. The float was filled with gasoline which could be released and filled with air at a moment's notice to send the craft bubbling to the surface.

The freighter, the Santa Maria, which carried the Trieste, was positioned about 250 miles southwest of Guam. The surface waves were very choppy as the Trieste was lowered from the safety of her mother ship, into the vast Pacific below. Slowly and silently the Trieste was allowed to drop; a free-fall that would take 4 hours and 48 minutes.

The Trieste creaked and moaned liked a rickety old chair supporting a 400-lb. man. Every knock and ping brought thoughts of disaster to the brave argonauts within. How great were the pressures? It is about the weight of an elephant standing on a postage stamp on every square inch of this steel-sided egg.

Picard passed the time looking at the small dials just inches from his face which would tell them that they had hit rock bottom. They were in constant radio contact with the Santa Maria positioned directly above but a universe away. In the event of a catastrophic failure they could offer little relief other than to bear witness to their crushing defeat.

Their meager mercury lights barely pierced the profound darkness except for an occasional sea creature passing by their window. Suddenly, a small window leading to the inside entrance cracked from the enormous pressures sending the occupants blood-pressure skyward. All that the pilots could hope was that it would hold.

Finally, a gentle thump signaled they had reached the bottom, some seven and a half miles below the surface. The crew stayed for about twenty nervous minutes, chewed some chocolate bars with chattering teeth, and returned to the surface. It was a chilly 45° Fahrenheit inside the cabin. They would take a number of important dial readings. The surface seemed to be oozing lava. This new lava spread the seafloor and was the mechanism causing the giant Tectonic Plates to move.

After twenty chilling minutes, the Trieste began its long three and a half-hour ascent back to the surface. Back on the surface an elated support crew welcomed their heroes back while the rest of the world took little note. The world was too gripped in a lunacy over the much anticipated moon landing.

We will be back as the seas hold an inexorable tug on our humanity. President Kennedy, an avid sailor, put it best in a speech in 1962 at the America's Cup. "I really don't know why it is that all of us are so committed to the sea, except I think it's because in addition to the fact that the sea changes, and the light changes, and ships change, it's because we all came from the sea. And it is an interesting biological fact that all of us have, in our veins the exact same percentage of salt in our blood that exists in the ocean, and, therefore, we have salt in our blood, in our sweat, in our tears.

We are tied to the ocean. And when we go back to the sea -- whether it is to sail or to watch it -- we are going back from whence we came."

33. Dinosaurs
Gideon Mantell Discovers Dinosaurs
1820

Dr. Gideon Mantell was a collector of things. He collected stamps, coins and curiosities of all kinds but in 1820 his latest odds and ends proved very odd indeed. In a short walk from her home in Sussex England, Mary Mantell, the doctor's wife, noticed an interesting rock in a gravel pit by the side of the road. Embedded in the stone is what appeared to be an enormous tooth? She showed it to her husband, who was immediately mesmerized by this strange curiosity. "What sort of creature could have such a tooth?" he wondered.

Iguanodon was the first dinosaur identified in 1820

Gideon looked for a match in all of the British museums but found nothing. Then, in a very strange and fortunate coincidence, he bumped into Samuel Stutchbury, who was a world authority on iguanas. Stutchbury immediately recognized the tooth as belonging to an iguana but confessed he had never seen any tooth that large. The two of them confirmed this first reaction by finding almost an identical match in Stutchbury's private iguana collection. The tooth was nearly identical but several times larger than any they had ever seen. Mantell returned to the site where the tooth was discovered and found several more fossil bones.

Mantell took a courageous step. Instead of just declaring it to be some unknown current species of iguana, He declared it to be an extinct ancient creature. He named it Iguanodon and the world had its first dinosaur. It would be up to Richard Owen to coin the term "dinosaur" in 1840, meaning "terrible lizard," but it was Mantell who would provide us with the first example. From the size of the tooth, Mantell estimated that the Iguanodon had to be about thirty feet long weighing about three tons. It was believed to have roamed the earth some 135 million years ago.

Soon hundreds of dinosaurs would be identified and they would be found in almost every corner of the globe. Strange bones that had been locked away in museums for years were now reexamined. These would include the Tyrannosaurus Rex or the "tyrant king," Stegosaurus, and the "three-horned" Triceratops. There was also the enormous flying Pterodactyl with a wing span larger than the Wright Brothers flyer. They found great sea creatures known as Ichthyosaurus that rivaled any nightmares that Columbus or his crew might conjure up. Today nearly 1000 species of dinosaurs have been identified and catalogued and we are still counting. Some are as tall as a five story building and others are as small as chickens.

A fight that could never happen - Dinosaurs died out 63 million years before man appeared on earth.

The last dinosaur died out some 65 million years ago but their reign lasted for about 160 million years. In a strange mystery known as the Great Dying all of the dinosaurs disappeared off the planet. Many scientists believe there is evidence that a large meteor hit the earth causing a huge amount of dust to spew into the air. This excessive dust blocked the sunlight killing plants that the dinosaurs ate and plunging temperatures on earth. We don't know for sure, but there is a large quantity of Iridium, a telltale element of meteors, found in sample rocks from this time.

In what is known as the Great Dying the dinosaurs disappeared 65 million years ago.

The eighteen hundreds were gripped by a dinosaur fever that continues today, all because of an unusual stroll by Mary Mantell. The next time you go for a leisurely stroll and kick a stone or a rock, take a closer look, it might just be the key to a bygone era, and you just might be taking a walk through Jurassic Park.

34. Einstein's Miracle Year
1905

As any high roller in Las Vegas will tell you, "hot streaks" are few and far between and when you are on a roll, it seems as though whatever you touch turns to gold. In 1905 Albert Einstein had a "hot streak" the likes of which have been unduplicated in scientific history. Einstein published three major papers that toppled the pillars of accepted science to make room for a new order.

The most important of the three papers was on Special Relativity oddly titled "On the Electrodynamics of Moving Bodies." In this seventeen page paper Einstein overturned our Newtonian view of time, space and simultaneity.

His second major paper was on the Photoelectric Effect, which sets the foundation for Quantum Mechanics, some of whose major tenets Einstein could never accept. Einstein tried to explain why shining low frequency light on a surface could never free the electrons to leave that surface. The energy from these low frequencies does not accumulate. One either supplies sufficient energy, a given "quanta," to the electrons to escape the atoms or one does not. It is like climbing stairs. One either climbs a whole stair at a time or one does not. One cannot climb half a stair.

His third paper of 1905 was on Brownian movement. It helped to solidify atomic theory. When tiny specs of grass seed or other particles are placed in a glass of water or beaker, they seem to get bounced around and move. Einstein explained this random bouncing as atoms colliding with the suspended particles.

**Albert Einstein
(1879-1953)**

These works would propel Einstein from an obscure Swiss patent clerk into one of the most well known physicists of the twentieth century. He once said, "To punish me for my contempt for authority, fate made me an authority myself."

35. Automobile – 1769

It is difficult to say who invented the first automobile or at what point the self-propelled "horseless carriage" became an automobile per se. Some place this milestone at 1769 with Nicolas Cugnot's "fardier a vapeur" or steam wagon. It was a three wheeled steam driven vehicle used for pulling cannons. It was made almost entirely out of wood and moved at a very unimpressive 2.5 mph. A leisurely walk is usually at the speed of about 4 mph. It also required large quantities of wood, and needed to be reloaded every fifteen minutes. This together with its forty-five minute start-up time made it less than optimal.

Cugnot's 1767 First Automobile was actually a tractor topping out at 2 ½ mph.

Cugnot's "fardier a vapeur" ran out of steam literally when in 1771 it hit a stone wall, counted as the very first automobile accident.

In 1876 German Nicholas Otto invented the 4-stroke gas powered engine and the car industry was kicked into high gear. Traveling on a train he noticed a smoking chimney "Of course, the engine must work something like a fireplace," he thought "with the hearth of the engine rich with all the necessary components for combustion." The key to his design was the compression stage. Now finally, we had a compact powerful source of energy. By 1891 cars were moving at an impressive 10 mph.

In 1908 Henry Ford is credited with bringing the automobile to the masses. In a time when cars were selling for over $3,000 Henry Ford offered a Model-T Ford for $850. You could

Ford's 1908 Model T sold 15,000,000 units. One could get it in any color you wanted as long as it was black.

get any color you wanted as long as it was black. Using his new "assembly line" approach to manufacturing he was able to crank out a new car every 24 seconds. He sold over 15 million in the first year.

Today the automobile is changing; hydrogen fuel cells and electric cars are gaining market shares by the day. Today there are over 230 million cars in the US. That is almost one car for every person. There are over 4 million miles of paved roads. What is not changing is America's love affair for the horseless carriage. It is the single most popular machine invented in the Industrial Revolution. It has provided a freedom unparalleled in American history and the call to the open road still beckons.

36. Sputnik
Soviets Launch First Satelite
October 4, 1957

US tried to launch its own satellite in 1957 which blew up on the launch pad. Newspapers dubbed it "Kaputnik"

On October 4, 1957 the Soviet Union launched the first man-made object into space. It was a 3 foot diameter satellite known as Sputnik. Americans flocked onto their porches to catch a glimpse of the new star which truly was too small to be visible to the naked eye.

Sputnik means "comrade traveler" when translated from Russian. But in 1957 the Soviet Union and the United States were anything but comrades. They were locked in a post World War II showdown known as the Cold War, cold since no one was shooting, at least not yet.

In Washington a panicked Lyndon Johnson screamed that soon the Soviets would be dropping bombs on us like rocks on a highway overpass. "I for one do not intend to go to sleep by the light of a communist moon!"

he ranted. Sputnik's gentle "beep, beep, beep" was the alarm clock to the American space program. It was clear that America was behind, and the leadership in Washington vowed to do something about it. America focused on space with new resolve, forming the National Aeronautics and Space Administration (NASA) to take up this challenge. President Eisenhower directed NASA to select astronauts out of 540 military test pilots. They were to be no older than 39 and had to posses a bachelor's degree. Out of this selection the first Mercury Seven astronauts were selected. They would soon lead us into space.

37. Mendel
Mendel Uncovers the Laws of Heredity - 1856

"An apple doesn't fall far from its tree." You've heard it said a million times recognizing the similarities between parents and their offspring. This we have known for a long time, but exactly how this is accomplished was a great mystery until an Austrian monk by the name of Gregory Mendel (1822-1884) began working in his garden in 1856.

Mendel found his answers in pea plants. He wondered how one could possibly get a white pea plant when one cross-pollinates two purple ones! Not only that, but why doesn't one ever get light purple when a purple plant and a white plant are crossed. After ten years and a thousand pea plants later, Mendel had his answer.

Gregory Mendel (1822-1884)

Being a clever mathematician, as well as a good gardener, Mendel noticed a three-to-one ratio between the numbers of purple flowers to the numbers of white flowers produced when the two are crossed.

Mendel proposed that each trait, such as color, had two parts. It received one of the two parts from each parent. If a flower received two purple genes, it became purple. If a flower received two white genes, it became white. If it received either a purple and a white gene, or a white and a purple gene, the purple would "dominate" the white to produce a purple flower. The white gene was called "recessive." He created a mathematical technique known as a Punnett square to keep track of these variables.

Mendel would test over 28,000 plants in a ten-year period until 1868 when he was promoted to abbot and he was forced to give up his investigation. Color was not the only trait Mendel researched. He would test several other "traits" including height, seed type, seed color, pod color and shape. All of them seemed to follow the statistical probabilities predicted by his Punnett Square.

Mendel's work was the beginning of modern genetics.

38. The Big Bang
Wilson and Penzias Detect Background Radiation
1965

In 1958 two talented astronomers, Robert Wilson and Arno Penzias joined Bell Labs at their new telescope in Holmdel, New Jersey. The big 20 foot ear shaped telescope resembled a giant conch shell one would use to hear the ocean. In 1965, this shell would hear the smoking gun of the Big Bang which started the universe some 13.7 billion years ago.

The idea for the Big Bang was first proposed by Georges Lemaitre in 1927. We knew from Edwin Hubble that the universe is expanding. If we roll the sequence back in time, that expansion would eventually converge into a single starting point and an initial explosion. It is a bit like chasing down the original rock that dropped in the water to make ripples in a pond.

When Penzias and Wilson began using their telescope they found that an annoying background radiation was always present. They assumed that it was coming from New York City

Big Bang occurred 13.7 billion years ago. The earth is about 4.5 billion years old

but when they pointed their telescope in that direction it did not get any louder. They then pointed at various points of the sky at different times of the year. To their amazement none of it made a difference. It was everywhere. It was as if they were inside this giant bubble of noise or radiation.

For scientists, looking out in space is like looking back in time. For example, when we look at the star Alpha Centauri, we are seeing it as it was 4.3 years ago not as it is today since it took 4.3 years for its sunlight to reach us. Some stars we see may have already vanished.

At the end of the visible universe Wilson and Penzias detected the huge background radiation which scientists believe is the remnants of the Big Bang. They had found the smoking gun of the Big Bang Theory and were therefore awarded the Nobel Prize in Physics in 1978.

39. Franklin's Kite
June 12, 1752

In the 1700's very little was known about electricity. Even the battery would not be invented until 1800 by Italian Alexander Volta. Printer, diplomat, author and inventor Benjamin Franklin set out to prove a very simple premise in 1752; that lightning and electricity were the same thing. He would fly a kite in a thunderstorm and see if he could get a spark to jump from a dangling key to his knuckle. It was not his intention to get struck by the lightning. Franklin had a healthy respect for electricity which had twice knocked him unconscious in the lab.

Menacing clouds darkened the Philadelphia skies on the afternoon of June 12, 1752. Benjamin Franklin raised his large head which was framed by his Quaker hat, "We have got to work fast" he said to his twenty-four year old son William. William was clutching a three foot triangular kite. A light rain began to speckle the ground.

**Benjamin Franklin
(1706-1790)**

Franklin worked feverishly to put the final knots on the kite's dangling tail. He liked the twine for this tail because its rough loose fibers stood on end when it was electrically charged. To the end of the connecting string he attached a large metal key. When the kite was charged, Franklin would then bring his knuckle close to the key and draw off a spark proving the presence of electricity.

The kite was composed of a large silk handkerchief and two cedar strips, which were made into a cross. On the top the kite sprouted a two-foot pointed antenna for attracting lightning. The changing air quickly sucked the triangular craft upward as planned and Franklin gave it plenty of slack. Soon the string grew taut trying to pull Franklin into the clouds.

Franklin never intended to get struck by lightning

The threads of the twine began to rise and stand on end. By now the twine and kite were both soaked with water. Franklin heard the crack of lightning and the rumble of thunder surrounded him. The flashing lightning could be seen reflecting off his father's bifocals, another one of Franklin's more famous inventions.

Franklin brought his knuckle close to the key, and the "little lightning" sparked the gap between the key and his finger. He laughed as the startling jolt threw him to the ground, a little shaken but not bad for a man of forty-six.

Franklin informed the world of his discovery. He was very diligent about writing to his fellow scientists and encouraged others to do the same. Since lightning was electricity he reasoned that it should be possible to draw the electricity along a preferred path rather than have it randomly striking a house or public building. This was a primary cause of many fires. This was Franklin's fundamental insight for the lightning rod. Soon it would pierce every sky from Philadelphia to Boston. He did not sell his invention but instead donated it to mankind as a gift.

Today thousands of buildings in every major city are adorned with Franklin's lightning rods. They have prevented countless fires. The Empire State Building alone gets struck several times a year since lightning always strikes the highest point.

Franklin's contributions to the electrical field were many. He would give us the positive and negative signs we still use today. He coined the terms battery, charge, conductor, and armature. Historian Turcot writes of him, "He tore from the skies lightning and from the tyrants their scepter." For a man that claims to speak for the common man, he was most uncommon.

40. Need a Lift? Otis Invents the Safety Elevator - 1853

Elisha Otis (1811-1861) did not invent the elevator. Elevators have been around since the time of the Egyptians and Greeks who used pulleys to lift large cargo. He did however; invent the safety elevator which made it safe for people to ride the elevator. In 1853, before Otis' invention, the elevator was considered a death trap. Workers refused to ride it or they had to be paid extra to work under those dangerous conditions. A simple snap of a rope would send the cargo and its passenger downward to an unpleasant finish.

While working in a bed spring warehouse, Otis designed a safety elevator. He attached a set of springs to the weight bearing rope. If the rope broke, the springs were automatically deployed sending a lever on each side to catch the tracks with teeth that ran the length of the shaft. It worked beautifully, but people still thought of elevators as dangerous devices and would not ride them.

In one of the great marketing ploys of all time Otis set up a demonstration at 1853 World's Fair in New York City. Otis constructed a four story open elevator inside the enormous dome of the Latting Observatory. He waited until a significant sized crowd gathered and then he started his next performance. Standing on the elevator as it rose, he explained that it was a new safety elevator and that no one could get hurt. When he reached the top of the great height, a man with an axe would dramatically cut the rope sending Otis and the elevator to what seemed an inevitable doom. The crowd gasped but the elevator slowed from a breakneck speed to a gentle stop.

"Gentlemen, I give you the safety elevator," Otis theatrically announced.

It was a trick worthy of Houdini himself and the crowds loved it. He would repeat this demonstration several times a day. Anyone who had any use for an elevator at all quickly made an order.

Elisha Otis would have little time to enjoy his success as he died shortly after this at the young age of forty-nine. His son Charles would take over the company and make improvements to the elevator to create a gentler ride. It's hard to imagine a skyscraper without the modern elevator. Instead of climbing a mountain of stairs, thanks to Elisha, the rise to the top is as easy as pushing a button.

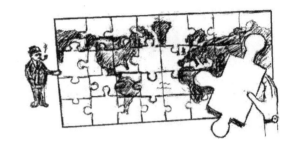

41. Pangaea
Wegener Discovers the World's Largest Puzzle - 1915

In 1912 German Alfred Wegener stared intently at a map of the world. He noticed how nicely Africa fit into the Gulf of Mexico. In fact, all of the continents look like they could fit together like matching puzzle pieces. Could they actually have been joined at one time?

Alfred Wegener (1880-1930)

He proposed that two hundred million years ago, all of the continents were connected in a one giant super continent called "Pangaea," which is the Greek word meaning "whole earth." Similarly, all of the oceans were likewise connected into one giant sea called "Oceanus." The remains of Oceanus is now the Pacific Ocean while the Atlantic came into existence much later. The continents slowly drifted apart until they came to their current positions.

No one paid much attention to Wegener's ideas for fifty years. The idea that the continents could move like crackers on the surface of soup seemed impossible for terra firma. But then evidence in support of his Continental Drift began emerging. Deep diving expeditions at the Marianas Trench revealed a seafloor that was spreading. Volcanic material oozed out of the seafloor. The continents were found to be floating on a subsurface ocean of hot liquid magma. The earth's crust is composed of several large continental size plates, tectonic plates, and several small ones float on this molten sea. The bumping of these plates causes earthquakes, volcanoes and mountain ranges. Wegener is now seen as a visionary. His ideas were slowly accepted and the scientific community eventually "drifted" in his direction.

42. Montgolfier Brothers – Sept. 19, 1783

Three unlikely traveling companions, a duck, a rooster, and a sheep fidgeted nervously in their large wicker basket. Above them was an enormous hot air balloon (1,400 cubic meters) built by two Frenchman, the Montgolfier brothers Ètienne (1745-99) and Joseph (1740-1810). The date was Sept. 19, 1783. A large curious crowd had gathered at the beautiful palace of Versailles just outside of Paris. Even King Louis XVI and Marie Antoinette were in attendance.

The original inspiration for the hot air balloon came from the Montgolfier's home fireplace in Annonay, France. Ètienne noticed how the sparks always carried upwards. "If only I could capture the force that makes these glowing sparks lift" the younger Ètienne wondered. He fashioned a small open bag out of silk and placed it over the fire, and "Voila," the bag rose. He called the lifting substance "Montgolfier Gas."

Brother Joseph unloosened the ropes, which tethered the balloon. The balloon lurched upward causing some ire to its unprepared threesome. The crowd gasped in utter amazement to see the balloon and the basket hovering free of gravity.

Marie Antoinette was enthralled. "Isn't this fantastic, Louis?!" King Louis XVI pinched his nose and waved off the offensive black smoke. The Montgolfiers incorrectly surmised that it was the black smoke, not the hot air, which provided the lift. As such they used fuel that would produce the most of this black smoke including wet straw, wool and even old shoes. One can hardly blame his Royal Highness for what truly amount to a royal stench.

Surprise gave way to a spontaneous applause as the balloon climbed higher and higher before finally reaching a height of over a quarter of a mile (500 m). The caged duck saw the size of his audience shrink by smaller and smaller percentages and he wondered how those people did that trick. The beautiful gardens of Versailles receded into a checkerboard of varied green squares as the balloon gained altitude. A casual bird kept a safe distance as it flew past.

Étienne smiled a bit and then began to worry about the balloon's landing. The previous August an unoccupied balloon was attacked by local farmers with pitchforks and scythes as it landed outside Paris. Luckily today no such incident took place. The balloon came bouncing down unharmed in a field about 2 miles away. The only casualty was a broken cage, which had previously contained a much-annoyed duck. The rooster waddled away, and the lamb took it in stride.

43. Tom Thumb
The Great Horse Race
Aug 28, 1830

The story was first recounted by John Latrobe (1866) about the first race between America's first locomotive, Tom Thumb, and a horse drawn stage. While demonstrating his new locomotive to officials of the Baltimore and Ohio Railroad, a horse drawn carriage from the renowned Stockton and Stokes Company was spoiling for a fight. They would race in an impromptu mythic struggle between a horse drawn carriage and the iron horse locomotive.

Peter Cooper was quite proud of his invention. It was called Tom Thumb and it was one of America's first steam locomotives. In the hopes of swaying the powers that be, that is, the Baltimore & Ohio Railroad, that it should be powered by steam, not horses, Peter Cooper had invited them out for a friendly demonstration.

It started as an uneventful day on the morning of August 28, 1830. The sides of the passenger car were flung open and the ladies grasped their hats as Tom Thumb steamed the train forward at an unheard of fifteen miles per hour. One man leaned back to soak in the sun and let the passing wind play havoc with his hair. It was a marvelous thirteen mile trip to Ellicott Mills, the turnaround point.

As chance would have it, a menacing cloud in the form of a gray horse pulled up parallel to Tom Thumb at the switching station. It was a magnificent stallion from the Stockton & Stokes Carriage Company, one of his main competitors for the rail line business. When the Baltimore and Ohio, B. & O. for short, had built the tracks, they had laid a second set parallel to the first.

The two parties exchanged pleasantries at an otherwise uncomfortable meeting between obvious rivals for the same contract. The driver of the Stockton & Stokes carriage challenged Tom Thumb to a thirteen-mile race back to Baltimore. Peter removed his top hat and rubbed the back of his neck with a handkerchief. He really didn't have a choice, he needed to accept.

In a few minutes both were ready.

"Go" yelled the starter as he dropped his arm and waved his hat to start the race. The great gray stallion exploded from the mark bursting out to a quarter mile lead. Tom Thumb's engine hissed, and snorted as it strained to get underway and build up steam. Steam engines were not known for their jack rabbit starts.

Within a few minutes Tom Thumb had pulled even with the horse and began to pull away. Then a belt which controls the airflow into the engine snapped and the engine struggled. The horse drawn carriage soon overtook Tom Thumb and high heeled it to victory.

Peter Cooper was finally able to fix the belt but it was too late for this race and the horse had won. But while the horse finished first that day, all that the crowd could talk about was the little engine that almost could. It was the beginning of the end for Stockton & Stokes and the "iron horse" would soon win victory after victory. Steam would soon replace horse power worldwide. The steam locomotive would become the dominant mode of transport for goods and materials for the next 100 years. The horse could not be reached for comment.

44. Pluto - 1930

Our on again off again romance with Pluto began in 1905. The scientific community noticed a slight wobble in the orbit of Uranus. Planets are not capricious; if it wobbles it is probably because there is a gravitational force pulling on it, perhaps another planet. By carefully reviewing the calculations scientists predicted exactly where this ninth planet should be. Industrialist turned astronomer Perceval Lowell would search for it for 30 years but the discovery of Pluto would not be his.

Perceval Lowell in 1905

It was left to 26 year old Clyde Tombaugh to spot an extra dot amongst a myriad of other dots in a series of night photographs to find Pluto. The year was 1930. Its discovery was special because it was the first planet found by calculation rather than by accidental observation.

It was an instant success. Walt Disney even named Mickey Mouse's new dog after the newfound planet. Also, the first two letters of Pluto are "P" and "L" which also stands for Perceval Lowell.

But was Pluto big enough to be a planet? It was, after all, smaller than our own moon. Second, was its orbit regular enough? Its orbit definitely didn't fit in the same plane with the other eight planets. Pluto's orbit is angled at an irregular seventeen degrees. Astronomers argued regularly whether it merited full planetary status.

The death knell for Pluto came when scientists found other Pluto sized objects beyond Pluto. Eris, discovered in 2003, had a distance at about 3 times the distance of Pluto and it was about the same size. Then there were many others beyond that, Santa, discovered on Christmas Day, Buffy and Sedna, just to name a few. Perhaps this was just too many planets for our tidy solar system picture.

The International Astronomical Union decided in 2006 to downgrade Pluto, and these others to a new category of dwarf-planets. Luckily Clyde Tombaugh, was not alive to see it. He died in 1997. His cremated ashes were sent on the space probe New Horizons which has a rendezvous with Pluto in 2015. A very upset public voiced their disapproval over Pluto's demotion. Bumper stickers began appearing "Honk if you love Pluto." Its status will be reviewed in 2019. Perhaps with enough barking Pluto might get its planetary status reinstated and it will be out of its intergalactic doghouse once and for all.

45. Galaxies
1926

In 1926 boxer turned astronomer Edwin Hubble, focused the newly built Mount Wilson telescope on a fuzzy star known as Andromeda. The enormous telescope bent its meager light and reflected it into 37 year old Hubble's eyes. He could not believe what he was seeing. What he thought was a single star was actually a beehive of individual stars, perhaps tens of thousands of them. He realized that the Milky Way was one of many similar galaxies and the universe was a lot bigger and more populated than anyone had first

According to Carl Sagan, "there are more stars in the sky then there are grains of sand on the earth."

Edwin Hubble (1889-1953)

supposed. In the end, the number of galaxies is estimated to be in the area of 500 billion, each with about 100 billion stars. This makes the number of stars in the universe about seventy sextillion, or seventy followed by twenty-one zeros. Carl Sagan would say it best in his book *Cosmos* "There are more stars in the sky then there are grains of sand on the all the beaches on earth."

46. Discovery of Radium
Marie and Pierre Curie Discover Radium - 1902

Marie Curie was fascinated by the recent work of Henri Becquerel. In 1896 Becquerel had placed some unexposed photographic plates in a closed drawer with some rock samples. To his amazement, the next day the photographic plates had been exposed. But since the drawer had been closed what was the source of light? He correctly concluded that light came from the rocks. A closer look revealed that it was the uranium that was emitting light rays, which Henry called Becquerel rays. He had discovered radioactivity.

Marie Curie (1867-1934) Her notebooks are still too radioactive to touch

Marie Curie wondered what other substances might give off these mysterious rays. She began by collecting samples of all kinds. Pierre, her new husband and senior by ten years, joined her in her work.

They soon found that pitchblende, a uranium ore, had a reading of about four times the strength of pure uranium. They began purifying the pitchblende to isolate the active substances.

The only place available at their Municipal School in Paris was an old shed previously used for dissecting. It was so miserable and dreary that even the medical school didn't want it. It leaked when it rained, sweltered in the summer and froze in the winter. One of Marie's diary entries reads that the temperature inside is a frosty 44° F, which Marie noted with ten exclamation points.

Slowly but surely they began to purify the mysterious substance. In June of 1898 the Curies identified the first of two new radioactive elements in the pitchblende. The first was "Polonium," named after her native Poland. It was 300 times more radioactive than pure uranium. But there was a second even more potent element than polonium. It would be called radium. It was nine-hundred times more powerful than uranium.

As it got more and more purified it began to glow. The Curies would return to their lab at night to see the sample mix glowing. As the radium became more purified, the glow was more intense, almost equal to a nightlight. It had a beautiful silvery and white color. Then horror! On the very last step of purification in 1902 the sample seemed to disappear. It seemed as though nothing was left in the sample plate. Where did it go? Then it dawned on them, it was there, but in such small quantities they could barely see it. The amount of the substance was minute, one millionth of one percent. It took several tons of pitchblende and four years of back breaking work to isolate only one gram of the substance. This is less than a pinch of salt.

In 1903 the Curies, along with Henri Becquerel, would be awarded the Nobel Prize in physics for their work in radiation. Marie became the first woman to receive this coveted prize. The Curies declined to attend due to ill health. Unbeknownst to them, both Curies were sick from radiation poisoning they received from their work. They were feeling exhausted, achy and had strange burns on their bodies. Pierre had a large sore in his chest where he happened to keep the sample of radium in his vest pocket. Marie's hands were badly burned, especially her fingers. Marie would eventually die from this radiation poisoning in 1934. Pierre's death came suddenly, not from radiation, but from a carriage accident on April 19, 1906.

Marie Curie (1867-1934) and Pierre Curie (1859-1906)

The Curies and the discovery of radium are a classic tale of scientific teamwork, perseverance, and hard work. They were tenacious in their pursuit of radium. Its isolation opened a whole new window to the atomic nucleus and helped to irradiate the future of science into the twentieth century.

47. "Camera Obscura"
Joseph Niépce Invents the Camera – 1826

Like the development of a photograph, the invention of the camera came in stages. Ibn Al-Haytham (965-1040) found that a hole put in a door or wall of a darkened room would produce an upside down image on the back wall of that room. This curious phenomenon was known as "camera obscura." Camera was the Italian name for "room" and "obscura" was the word for dark. Artists used it for a thousand years to help trace the proper geometric proportions of the landscapes that they were painting. Unfortunately, there was no way of saving this inverted image other than tracing it.

In 1826 a Frenchman named Joseph Niépce realized that one doesn't need a whole room, just a small handheld box with a hole in one side and a back side to the box and it will accomplish the same thing. This was the first inspiration for the modern camera.

Although he had miniaturized the "camera obscura," Niépce still needed a way to save the image. He knew that it was possible. Fabrics left in the light for a long time faded. Perhaps in the same way he could get his inverted images to "fade" into a material that was light sensitive on the back wall of his box. Niépce heard about Johann Schultz who was able to burn stenciled silhouettes onto a bottle filled with silver chloride. In 1816, Niépce tried coating the back of the camera with a piece of paper covered in this same solution. To his amazement, an image appeared on the paper. There was however, one slight problem, the light areas and the dark areas were reversed! He had created a negative image. He was one step away.

Finally, in 1822, Niépce took the "negative" image and used it as the input to the camera, reversing the lights and darks one more time and his picture came out. He also replaced the silver chloride with an asphalt coating used in lithography and this also worked.

Unfortunately, his images faded away after just a few hours. He reasoned that if he had a stronger light or increased the exposure time, perhaps the images would last longer. He stuck his camera lens on the window facing his backyard. He allowed the brilliant light from the sun to burn an image for eight straight hours. The result is what is considered the world's first photograph. It was a picture of the courtyard and surrounding buildings at his country villa outside of Paris.

The fast talking Frenchman Louis Daguerre talked Niépce into a partnership. Daguerre was a master promoter and showman. He owned the famous Diorama, an exhibition hall in Paris. Unfortunately Niépce died shortly thereafter in 1833 and Daguerre stole most of the credit for the invention of photography. In fact, the first photographs were known worldwide as "Daguerreotypes" and he sold millions.

Taking pictures became a world wide obsession. In 1888 George Eastman did for the camera what Henry Ford did with his Model-T, he built a camera for the masses. For twenty-five dollars one got the camera, film for 100 pictures, and free film processing. America and the rest of the world went camera crazy.

Today's pictures are created in a variety of forms including digital cameras which store images in millions of pixels seared into computer chips. The process has changed but the original dream, to freeze a fleeting moment in time forever has not. It is difficult to put into a few words, the effects of the camera in our lives; wedding pictures, baby pictures, pictures of a lost loved one, the Wright Brother's first flight, Mathew Brady's chronicle of the Civil War. If a picture is worth a thousand words one cannot say enough about Joseph Niépce's invention, a man who like his early photographs, has now faded into "camera obscurity."

48. Dynamite
Alfred Nobel Tames Nitroglycerin – 1867

From its discovery nitroglycerin, the active ingredient in dynamite has always had a very bad temper. It is a trait first learned the hard way by its discoverer Ascanio Sobrero of Italy. In 1846 Sobrero had just isolated the substance in his laboratory when an errant breeze exploded his sample blowing up his laboratory injuring many including Ascanio.

But the world had a need for a stronger explosive. Black powder simply lacked the punch for building railroads through mountains and removing millions of tons of earth for canals. Nitroglycerin was five times more powerful than black powder but no one dared use it. Alfred Nobel, one of Ascanio Sobrero's students, was determined to tame this lion.

In 1867 while transporting a truck load of nitroglycerin canisters in Hamburg, Germany, a leak developed and the nitroglycerin poured into the soft packing sand. This sand was called "Kieselguhr."

"Why had the cargo not blown up?" Nobel wondered. Here finally was Nobel's answer, mixing it with this soft sand had rendered the nitroglycerin harmless. He packaged it in small cylinders of wax paper and called it dynamite after the Greek word "dynamis" which means power.

This new "dynamite" was now incredibly safe compared to its former state. Nobel could drop it; crush it, even light it on fire and it remained dormant. In fact, the only way to explode it was with a blasting cap and fuse which he also invented. It was a great combination and an instant success. In the first year alone he sold over eleven tons and within ten years this grew to 65,000 tons. He became one of the richest men in the world with over 90 plants in 20 countries.

90% of Mount Rushmore was carved by dynamite

Still, the dark side of dynamite cannot be minimized. Nobel even lost his brother when static electricity blew up the plant he was working in. It was also soon apparent that his dynamite could not only blow up mountainsides but people as well. It quickly became a weapon of war. It was the original weapon of mass destruction and it killed millions during the next two World Wars.

Alfred Nobel argued that dynamite was neither good nor bad. He felt that like a knife, the goodness or badness of the instrument lies in the hearts of men not in the tool itself. He would later have terrible remorse for his invention and established the Nobel Prizes in an altruistic gesture.

It is hard to imagine how long it would have taken to build the Transcontinental Railroad or the Panama Canal without dynamite. It revolutionized the mining industry. The Hoover Dam alone required more than 4 million pounds of it. Ninety percent of the faces of Mount Rushmore were carved by precision dynamite.

In a strange twist of irony that Nobel later in life developed a painful heart condition. The only medicine with which to alleviate his symptoms was none other than nitroglycerin.

49. Under the Pole
Nautilus Confirms No Land under the North Pole
August 3, 1958

What is the difference between the North Pole and the South Pole? For one thing penguins are found only on the South Pole. But also there is no land under the North Pole. If one looks at a model of the globe there isn't any land mass representing the North Pole. There is one for the South Pole, Antarctica, but nothing but an ice cap for the North Pole. Of course, no one could confirm this until August 3, 1958 when the atomic submarine the USS Nautilus passed silently underneath it.

It was a milestone quietly marked by Captain William Anderson, "For the world, our country and the Navy – the North Pole." The date was August 3, 1958.

The Nautilus was unique in that it was powered by a small nuclear reactor and thus could stay submerged for long periods of time required by the exploration. It had submerged at Point Barrow, Alaska on August 1, 1958 and would reemerge in the Greenland Sea some 96 hours and 1,590 nautical miles later. It was her third attempt.

One year later, in 1959 the submarine Skate recreated the trip but this time surfaced at the Pole. Its reinforced steel hull broke through the frozen

20-foot ice exactly at the North Pole. A few precious rays of the weak Arctic sun shed light on one of science's great unanswered questions.

51. Black Holes
Schwarzschild Finds a Bizarre Solution to Einstein's Field Equations – 1916

Karl Schwarzschild (1873-1916) lay dying in a hospital bed just a few miles from the where he had gotten sick. The year was 1916 and Germany was in the midst of WWI. The German Lieutenant had contracted a horrible skin disease in the trenches of the Russian front. There was no known cure for it. He wondered if Einstein had read his paper on an extraordinary consequence of Einstein's field equations, a black hole.

Schwarzschild detailed a radius around a sun, known as the "Schwarzschild Radius" which is the break-even point between the speed of light and the gravitational attraction of that sun. If one gets any closer, they are doomed, inexorably attracted to the center, or singularity, and crushed beyond recognition into one giant atomic nucleus about the size of a basketball. It is so concentrated that one teaspoon of this nucleus weighs 100 million tons or about the weight of one thousand tractor trailers. In fact the escape speed of such a sun would exceed the speed of light. Since no light could ever leave such a surface to then strike our eyes, it is black, or devoid of radiating light. In other words, it is a black hole.

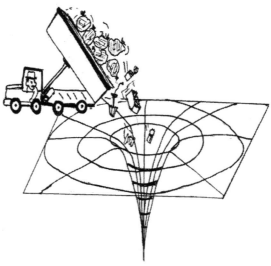

Black Holes are the ultimate trash compactor. Everything gets crushed. One teaspoon of its nucleus weighs as much as ten thousand loaded tractor trailers

Einstein presented Schwarzschild's work on Schwarzschild's behalf to the Prussian Academy in 1917 and today it is marveled at as a masterpiece. New solutions were found, some of which included a gray hole and even a white hole as well.

We believe we have identified at least one such black hole in Cygnus-X in 1985. Stephen Hawking would gain fame as the physicist who would detail this strange new world. Schwarzschild would never know of Einstein's response. In 1916, his eyes grew heavy, and like the phenomenon he studied, the darkness closed in. He died at the early age of forty-one, far too early for one of astronomy's brightest stars.

52. Edison's Favorite
November 20, 1877

Edison marketed the first talking doll

Often a great invention is discovered while pursuing another. Such was the case with Edison's phonograph. Edison had actually been working on improving the telegraph when he noticed that his voice made small indentures into a piece of paper in back of the speaking cone. When he ran these bumps underneath a needle he found that it had captured the source of the sound. He tried it with tin foil that was a bit more hardy than paper.

"Helloooo!" bellowed Thomas Edison into what looked like a large black ice-cream cone. He slowly turned the crank, which moved a needle at the end of the cone into a new position on the tinfoil cylinder. He repositioned the needle back to the beginning of the tinfoil and the machine responded in a metallic, "Helloooo!" Edison chuckled and continued.

"Mary had a little lamb, its fleece was white as snow, and everywhere that Mary went the lamb was sure to go," Edison recited. Placing the needle back to its starting point, the tinfoil faithfully repeated Edison's rhyme.

Within two months he was ready to patent. Clerks at the United States Patent Office could hardly believe their ears. They thought Edison was pulling some sort of joke or parlor trick. They had never seen or heard a machine that talked, nor had anyone else.

The phonograph soon swept the nation. Edison's talking machine was in every amusement park and the public clamored for more. For five-cents one could hear two minutes of music. Edison even made the first talking doll, although it proved unreliable and was a marketing failure.

Within a year, Alexander Graham Bell improved the tin-foil cylinder to wax. Later in 1888 German inventor Emile Berlinger changed the cylinder shape to its more familiar flat disks or pizza shape. Edison's recordings had to be made one at a time, while Berliner's could be stamped out one at a time in seconds and mass produced by the millions. The compact disc did not arrive until 1982 and the IPOD came in 2001.

Today music is spliced and diced, digitized and synthesized but it was Thomas Edison who first recorded the human voice. It's hard to imagine a world without it. At the flick of a switch we can bring the entire Boston Symphony into our living room or head phones. We can waltz, break-dance, or hip-hop anytime day or night. Of all of Edison's 1098 patented inventions this was his favorite.

53. Galileo's Freefall Experiment 1590

Tower of Pisa is predicted to fall in the next 100 years

Aristotle theorized that heavier objects would fall faster than lighter ones. Galileo believed that he was mistaken. Galileo hypothesized that air resistance is what made a feather and a cannon ball fall at different speeds. Devoid of air, the two objects would fall at the same rate.

The twenty-six year old Galileo eagerly climbed the eight stories of the leaning tower of Pisa in 1590. In his possession, Galileo had two balls, one weighing ten pounds and one weighing one pound. Galileo peered down the tower and equally positioned a ball in each hand and let them fly.

It was tough to call, but the cannon ball hit a split second before the smaller ball. Galileo knew that the cannon ball would hit first, but only very slightly. He recalls in his book, *Dialogues Concerning Two New Sciences* (1638) "You find, on making the experiment that the larger outstrips the smaller by two finger breadths …now you would not hide behind these two fingers."

Galileo's proof would not come for another 400 years until the Apollo 15 lunar mission. On August 1, 1971 the last day of Apollo 15, astronaut David Scott simultaneously dropped a hammer and a feather on the surface of the moon. As one knows the moon has no appreciable atmosphere and therefore no air resistance. To no one's amazement, as Galileo had predicted, without air resistance, both objects fell in equal time. It was a great moment of science and a feather in Galileo's cap.

54. Small Pox Eradicated from the Planet December 1977

Small Pox is a particularly deadly disease. It is estimated that it killed over 400 million people in the 20th century alone. Fifty percent of its victims died while those that survived carried horrible scars to mark their struggle.

In 1797 Dr. Edward Jenner of England developed the first vaccine against smallpox. In the one hundred years that followed, as vaccination became widespread, smallpox cases dropped dramatically dropped from forty out of every ten thousand in 1800 to one in every ten thousand by 1900. Still in 1967 the number of smallpox cases even at this ratio was still staggering, fifteen million worldwide including two million deaths.

Only small amounts of the small pox virus remain. They are kept tightly locked at several disease centers around the world

In 1967 The World Health Organization (WHO) decided to act decisively. Spearheaded by Epidemiologist Donald Ainslie Henderson they hunted out and destroyed any outbreaks of this dreaded disease around the world. The selected medical teams would wait, perpetually vigilant, bags packed, for the next outbreak to occur. They would then move in to immunize, isolate and educate the infected community.

By the 1970's smallpox had been cornered to just a few isolated spots in Africa and India. Finally in 1977 the last case of smallpox was reported in Somalia and treated. Four decades later, humanity has declared a cautious victory over this vicious killer. To date, it is the only infectious disease to have been eradicated from the planet. Small amounts of smallpox samples are kept carefully quarantined at the Center for Disease Control (CDC) in Atlanta, Georgia. Other sites include London, Moscow and China. They have been kept for study. Some argue the risk of this plague escaping is far too great and that these final stockpiles too should be destroyed

55. Finding the Titanic
July 13, 1986

On the night of April 14, 1912 the majestic Titanic, the world's largest ship, struck an iceberg off the coast of Newfoundland and slowly sunk into the icy North Atlantic. The Titanic took with her 1,513 passengers to a watery grave. Another 706 survived in the half-filled lifeboats that escaped in the confusion of that night. The Titanic became the most famous shipwreck in history.

On September 1, 1985 Dr. Robert Ballard and his crew aboard the U. S. Knorr spent twenty-two days searching for the missing Titanic with sophisticated sonar equipment. In a method some refer to as "mowing the lawn" the ship would trace back and forth mapping the sea floor. A crewmember watching a monitor of the submerged camera screamed out "Wreckage! We have wreckage here."

The rest of the crew hurried to the monitor in what they hoped would be a momentous discovery. Wreckage, yes, but was it wreckage from the Titanic? They waited for some signature piece, something that would uniquely identify this wreckage as Titanic. The two-mile deep cameras would provide that affirmation.

Robert Ballard (1942-   )

"Boiler!" The call came out. This was it. It was in fact the Titanic. Titanic's boilers were legendary and unmistakable. They were enormous, each about the size of a three story building and here they were.

On July 13, 1986, Ballard of Woods Hole returned to the wreckage site. He hoped to be the first visitor to the great ship in 74 years. Aboard the Atlantis II, he was armed with a small submarine called the Alvin and its robotic dog Jason Junior.

The Titanic was positioned 350 miles east of Nova Scotia and it was about 1,000 miles from Ballard's port of sail, Woods Hole, Massachusetts. Ballard and his crew entered the small submarine Alvin, which would take them the 2½ miles down to the murky bottom.

After two and a half hours they reached the wreck. The Jason Junior, the remote robot vehicle motored out of the back of the Alvin for a closer scrutiny. Jason was about the size of a car trunk. It would be used because of the dangers inherent in a rotting ship.

The television cameras aboard the robot recorded some stunning images of this ghostly ship. The first surprise was that the ship was broken in half with the bow and the stern separated by some 1,800 feet. This confirms some eyewitness accounts that saw the ship break in two just before disappearing beneath the surface. Ballard reasoned that for the two pieces to be that far apart, they would have needed to start drifting apart at the surface. One of the halves was actually turned around and facing the opposite direction.

Some thought was given to raising the Titanic using thousands of ping pong balls. But the ship was also in bad shape, far worse than anyone expected. It was embedded in the sea bottom it occupies, sunk up to its anchors in mud. The natural processes of the deep had taken their toll on this colossus and it was now a habitat for curious fish, coral and countless forms of life. The Atlantis II would continue to explore for 9 days. Upon departing, a commemorative plaque honoring the Titanic dead was left at the site.

Before the exploration of the Titanic, we once considered anything lost or buried at sea as gone forever. The weight of the water pressure and the lack of visibility and mobility made it next to impossible to go more than a few hundred meters below the surface. The 1986 exploration of the Titanic changed that. The magnificent ship, not seen since the icy night of April 14, 1916 was seen again, no longer buried deep below the surface of the ocean or our consciousness. As for Ballard's next project, he will be looking for Noah's Ark.

56. A Most Unusual Experiment
1794

During the French Revolution and the Reign of Terror that lasted from 1793-1794 many noblemen and others met with their bloody deaths at the end of the guillotine's blade. Stories circulated about decapitated heads still being conscious after being severed. One woman's head was lifted to the crowd, and showed displeasure when the crowd applauded her demise. In another case, two bitter rivals had both their heads fall into the same basket where one foe bit the other.

One such doomed scientist agreed with his aide to make the best of a bad fate and test this theory that one survives for a certain length of time after beheading. The aide counted 15 blinks at a blink per second. It is believed that it was the great Parisian chemist Antoine Lavosier. Lavosier was one of the founding fathers of modern chemistry but he was a tax collector which made him quite unpopular with the masses. Mathematician LaGrange laments of Lavosier's death "It took them only an instant to cut off that head, but France may not produce another like it in a century."

None of the official biographies of Lavosier mention this experiment, but if true, he was one scientist who kept his head about him even while losing it.

57. Dwarf Planets - Oct. 31, 2003

Xena - The Warrior Princess

It has been cataloged as UB313 which is short for "unidentified body 313." The discoverers Mike Brown and Chad Trujillo have named it Xena in 2003 after the main character of a popular television series called "Xena, the Warrior Princess."

Xena is just a tad bigger than Pluto. Pluto has a width of about 1,470 miles across while Xena's diameter approaches 1,860 miles. It orbits the sun at about 3 times the distance of Pluto.

The question before the International Astronomical Union (IAU) is whether Xena should be given planetary status and be the 10th planet. Pluto for example is smaller than our own moon, which is 2,160 miles wide.

In September of 2006 the IAU met in Prague. It was decided that Xena would not be a 10th planet. As of 2006 its new name is Eris, instead of Xena and it would be categorized as a new class known as "dwarf planets."

Perhaps the Coup de Gras for Xena and Pluto was the looming dozen or so sizeable objects now suspected to lie at very great distances beyond Pluto in what is called the Kuiper Belt. Are we going to grant them all planetary status? Amongst those waiting on the list for planetary status are the objects Sedna, Santa, Buffy and Ixion.

In elementary school models of the solar system have always been among the favorite choice for science projects. The planets are usually

Eris, formerly Xena, is a tad bigger than Pluto. Xena's diameter is 1,860 miles while Pluto tops 1,470 miles.

represented by balls of varying dimensions of Styrofoam. If all of these other bodies are to be included as new planets, someone needs to get the word out that in this year's models they are going to need a lot more Styrofoam.

58. Brave Heart
The First Heart Transplant - Dec 3, 1967

Dr. Barnard performed the first heart transplant on Dec. 3, 1967

South African Louis Washkansky was the original man of steel. He served two tours of active duty during World War II one in North Africa and then later in the Italian theatre. Active his whole life, at 54 Louis was no longer indestructible. In fact, after three serious heart attacks, he was close to death. Before departing this world, however, Louis would make one last brave decision by consenting to be the world's first heart transplant recipient. Louis knew that as such a first, his odds were not good to survive very long, but Louis saw a chance to do something bigger than himself and he agreed to give it a try.

Dr. Christian Barnard, a young surgeon from South Africa would perform the nine hour operation with help from 30 medical people. The donor of the heart was Mrs. Denise Darvall who had recently died tragically in a car accident. The operation took place on Dec. 3, 1967 at the Groote Schuur Hospital in Cape Town, South Africa. The operation was a success but soon afterwards Louis Washkansky's own immune system began attacking the transplanted heart as a foreign invader. Large doses of drugs had to be administered to surpress his immune system's rejection. In the end,

Louis Washkansky survived only 18 more days before succumbing to pneumonia.

A second transplant operation was tried in 1968 on Philip Blaiberg. Phillip lived a considerably longer amount of time, 19 months to be exact. In 1969 Mrs. Dorothy Fischer survived an amazing 24 years after her heart transplant proving that such surgery could have long term benefits. Today in the United States there are over 2200 heart transplants a year. It is the third most common transplant in the country. (Corneas and kidneys are the first).

Thanks to the hard work of Dr. Barnard and the bravery of Louis Washkansky many patients around the world are able to hear a few more beats to this symphony we call life.

59. Bubble Wrap
Pathfinder Makes a Unique Landing on Mars
July 4, 1997

There have been many unmanned space probes, why is this one unique enough to be considered one of the great moments of science? It was quite simply one of the craziest ideas ever invented and it actually worked. It was wrapped in a sort of high tech bubble wrap and came bouncing to the Martian surface like an errant basketball.

A new star appeared in the Martian sky on the morning of July 4, 1997. It burned a brilliant yellow as its heat shield shredded the salmon sky and friction slowed its descent. As it grew closer, it became clear this was no ordinary meteor but a hunk of metal launched by NASA on December 4, 1996 and 100 million miles away. It was the unmanned Pathfinder with a cargo of a 2 foot rover Sojourner inside. With 7 miles to go a parachute unfurled to slow its descent amongst the ¼ gravity of Mars. With about 1000 feet to go a dozen or so large canvas balloons inflated to form a sort of protective bubble wrap to cushion its fall estimated to be at a modest 22 mph. The expected first bounce carried it skyward 50 ft high. It bounced once again this time 23 feet high, and then again and again. It bounced for a total of 92 seconds before coming to rest on the Ares Vales flood plain with a frigid minus 64 degrees Fahrenheit.

The command was given to deflate the protective airbags, but at a distance of over 300 million miles, each command from NASA would take over ten minutes, even at the speed of light. When the balloons deflated its precious cargo, the Martian rover called Sojourner was released. The rectangular rover measured no more than one foot by two feet, and was less than three feet tall. Despite is size, its 2-inch wheels had big shoes to fill. It was the first probe to visit Mars in twenty-one years. It would end its mission in September having returned over 2.3 billion bits of information and 550 pictures. It would also sniff the soil and do some important chemical analysis.

Back on earth, the pictures from the rover were nothing short of spectacular. No sign of life, at least macroscopic, despite it being equipped with three cameras, two in front and one in the stern. Its lunch tray back was covered with solar panels to drink the meager sun's rays which graced the surface.

Because of the distances to Mars, it would take 10 minutes for the rover to respond to signals.

Pathfinder was an incredibly well thought out enterprise conducted for the bargain basement cost of $197 million, just a trifle to what a manned mission to Mars would be. The cost for that is estimated at $400 billion. The Apollo missions by contrast cost $170 billion. "Better, faster, cheaper" was NASA's new motto under director Daniel Goldin's vision. Three cheers for NASA for a mission with a little bounce.

60. Color Coding
Gram Finds a Way to Stain Bacteria
1876

Every college student knows the value of highlighting and color coding to isolate the important facts in a pile of textbooks. In 1876 Danish bacteriologist Hans Christian Gram gave us a way to color code bacteria. He found that bacterium with a thick cell wall, such as Streptococcus pneumonia, stain purple with dye. This is known as Gram-positive. Bacterium with a thin cell wall, such as E. coli, stain pink and are known as Gram-negative. This simple color coding is important in identifying and treating bacterial infections. For example, Gram-negative bacteria are fairly resistant to most antibiotics while Gram-positive bacteria are responsive to them.

61. "Eureka!"
Archimedes Uncovers Jewel Fraud (247 BC)

"Fraud!" King Hiero II screamed in 247 BC. The monarch of ancient Syracuse felt that he had been swindled. He had given some untrustworthy jewelers an amount of gold to craft a new crown for him. When the jewelers were finished, King Hiero II was under whelmed by the new crown suspecting that it was not the pure gold as he had contracted. But how could he prove it without destroying the crown? He turned over the problem to resident genius Archimedes (287-212 BC).

Archimedes was deeply puzzled by the problem. That night he decided to take a bath to help him relax. As he sank into the warm tub, his inspiration and the water overflowed.

"Eureka!" he screamed "I have found it!"

Archimedes was so excited he left the bath tub and forgot to put on his robe. He ran through the town naked.

Archimedes gathered the King and the two anxious jewelers into a great hall. He took the questionable crown and immersed it in a water bath filled to the brim. The crown displaced some of the water, which Archimedes carefully weighed.

"I'm afraid Sire, that the crown is not pure gold."

The outraged jewelers protested "What kind of foolishness is this?"

Archimedes explained how the spilt water represents the exact volume of the crown. Its mass was also easily determined. With these two pieces of information he was able to tell the density of the crown, a telltale identifier of a substance. A simple calculation showed the density of this crown was inconsistent with the known density of gold.

The King clapped to show his approval. The two disreputable jewelers were less impressed. They were taken away and placed in another metal – iron, in fact, one set of bracelets and one set of anklets, not of their choosing. King Hiero II had his moment of truth, a good man is hard to find, he reflected, and a scientist like Archimedes is worth his weight in gold.

62. Elusive Ether?
Michelson and Morley Dispels the Ether Theory – 1887

What then is gravity? It is a mystery that Newton's law of gravitation failed to answer. It accurately described the force's strength and variables but stopped short of an explanation as to what it actually is.

What was the mechanism for gravity? One popular idea was Descartes vortex theory. When one pulls the plug on a sink the draining water forms a whirlpool or vortex sucking in anything floating on the surface. In 1644, Descartes used this

Descartes whirlpool theory of gravity - instead of water space is filled with ether

mechanical analogy to explain gravity. The planets orbited the sun in much the same way as debris orbits the center of a whirlpool being pulled into the middle. But where was the water?

Certainly space was not filled with water. Descartes replaced it with a fuzzy sort of "ether." Ether was a vague invisible sort of stuff that was the medium for this solar system whirlpool. It was an appealing idea to many.

Was there an "ether"? In a critical experiment by two of the greatest experimenters of the modern era, Albert Michelson and Edward Morley sought to find out. They had gained fame and a Nobel Prize in 1926 for accurately measuring the speed of light.

In 1887 they reasoned that if the earth were riding this ether around the sun, the ether must have a direction. Light traveling with the ether would travel faster and light traveling against the ether would travel slower.

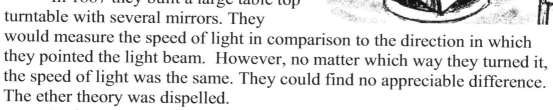

In 1887 they built a large table top turntable with several mirrors. They would measure the speed of light in comparison to the direction in which they pointed the light beam. However, no matter which way they turned it, the speed of light was the same. They could find no appreciable difference. The ether theory was dispelled.

Today gravity still remains a deep mystery, one of the four forces of nature. Despite this lack of insight, no one has tried to resurrect the specter of an ether. Gravity seems to act in a void without a mechanical medium, an unavoidable conclusion that has left some in science feeling a little empty.

63. Oxygen
Joseph Priestly Discovers Oxygen
August 1, 1774

When a person is deprived of air he or she dies, that much we knew. But what exactly constitutes air was not understood until the time of the American Revolution. On August 1, 1774 a Unitarian minister and British chemist Joseph Priestly (1733-1804) identified the key element, which is oxygen.

Priestly unwrapped his new magnifying glass, which he had just received from a friend. "No more candles," he thought "This will make his study of gases much easier because he could heat a substance in a bell jar without lifting it up.

Priestly couldn't wait to try it out. He placed some red mercuric oxide powder underneath the bell jar and then focused his lens. A tube led from the jar into a pan of mercury with an upside down glass in it. As gas was produced, the new gas would go through the tube and push out the mercury showing a space at the top of the overturned glass.

His new lens worked just as well as he had hoped and a measurable amount of gas was trapped in his overturned bottle. When he turned the bottle upright the candles in the room flickered with brightness not seen before. Was there a connection?

He took another candle and blew it out. He then placed it quickly inside the bottle containing the trapped gas. To his amazement the still hot and smoking wick roared back into a full burn. Priestly had discovered oxygen.

He would go on to find that it composed about 20% of air. The other 79% would be nitrogen, which gives the sky its beautiful blue color. Xenon, krypton, argon and a host of other elements make up the remaining 1%.

Priestly would not stay in his Sherborne home long. His sympathies for the French Revolution made him a hated man. In 1791 a mob looted and burned his home. He immigrated to America in 1794 settling in Pennsylvania where he spent the rest of his life. He found the air very tolerant for those with a revolutionary spirit, and there he breathed a little easier until his death in 1804.

64. Exo-Planets - 1988

What exactly is an exo-planet? It is a planet not belonging to our solar system. In particular it is a planet that orbits around another sun or star. One will recall that a star is nothing but a sun very far away.

In 1961 Frank Drake, a University of California, Santa Cruz professor, came up with a Drake's Equation to try and guess the number of planets in the Milky Way galaxy that are capable of sustaining life. By his factors there are a possible 20 billion planets capable of sustaining life in just our galaxy alone.

He took the number of stars in the Milky Way, about 400 billion, and estimated that perhaps 1/3 of these suns have planets orbiting about them, or 133 billion solar systems. If each of these solar systems had about 10 planets each, that means that just in our own Milky Way galaxy alone, there are a possible 1,300 billion planets. How many of these planets are earth-like and may have developed life reduces this number to 100 billion worlds.

It was an amazing guesstimate. But up until 1998, we could only guess that another sun had planets. In 1988 we had the first evidence of a planet orbiting another sun found by Canadian astronomers Bruce Campbell, G. Walker and S. Yang. It was a gaseous giant planet orbiting gamma Cephi (also known as Alrai). This finding was later confirmed by other astronomers. We can't actually see the planet because our telescopes are not that strong, what we see are its gravitational effects on its host sun or other celestial bodies.

This time make the crop circles go counter clockwise that will really bring in the crowds

Since 1998, the planet search game has gone crazy. As of November of 2010 we have evidence of over 400 exo-planets orbiting various stars and new ones are added at about two dozen every year. Of the 400 exo-planets found, none have been earthlike yet. Most have been gaseous giants like Jupiter.

Does this mean that there are very few earthlike planets? No, we are seeing the gaseous giants simply because they are bigger. Earthlike planets are smaller and therefore more difficult to find. They are just beyond our ability to see with our current instruments but we believe they are there. A very large telescope, the Magellan, ten times more powerful than the Hubble telescope is now being built. Perhaps then we may make direct observations of these exo-planets.

There are a growing number of "rare Earth" supporters that think our conditions for life are quite rare, that in the cosmic game of dice for habitable planets we are the only one to hit the jackpot. Most scientists believe that if the conditions for life are present, in terms of molecules, and temperatures, life will form. Life, once it begins, is tenacious, and it will grow, mutate, and evolve. With this perspective, if one grants the simplest amoeba, one then conceded the possibility of plants and animals. If one grants simple plants and animals they will eventually evolve into higher creatures. Is it time to roll out the welcome wagon for ET? Do we need to brush up on our Andromedean etiquette?

65. La Brea Tar Pits - 1901

In the area which is now the city of Los Angeles, at the foot of the Santa Monica Mountains was a large plain known as the La Brea Tar Pits. From a distance the shallow black tar pits looked like pools of water as they reflected the blue sky above. Thirsty birds and animals approached the pits hoping to get a cool drink of water and a moment's respite from the hot California sun. To their dismay, instead of water they discovered hot tar oozing from the small holes. Most managed to escape, but some were not as lucky and became trapped in the gooey asphalt. Some simply died of heat exhaustion. Others starved to death over time. Still, others fell prey to hungry beasts. Predators seeing a quick and easy meal moved in for the kill only to find themselves victim to the same trap.

This cycle continued year after year, millennium after millennium, for over 40,000 years. The result is a mind boggling collection of fossilized animal bones and plants. The count is a staggering one million plus specimens and still counting. There are over 2,000 saber-tooth tigers alone, just to name one species.

This amazing find came to light in 1901. Workers at the asphalt factory used dynamite to unearth some commercial grade asphalt that lay beneath the surface. These explosions caused previously hidden layers to surface. Factory workers and paleontologists noticed large numbers of bones imbedded in the asphalt and the excavations began.

To date, over 231 species of vertebrates, 159 kinds of plants, and 234 invertebrates have been identified. The most common are the dire wolves, which total several thousand. The collection also includes bison, horses, wolves, turkeys, mastodons and even fourteen human skeletons. The distinction between hunter and hunted is tenuous at best. Nature can quickly turn the tables and an easy meal can quickly change to disaster. It is a grim reminder that nature is unforgiving and life and especially death can be "the pits."

66. Sound Barrier
Charles Yeager Exceeds Mach I - October 14, 1947

The wind whistled and pinged through the helmet of Charles "Chuck" Yeager (1923-) as he climbed into the small X-1 plane. It was an experimental plane built by Bell labs to try and break the sound barrier, or Mach 1, a barrier that had seemed impenetrable. The X-1 had been carried aloft over the skies of Edwards Air Force Base in California. It had been named "Glamorous Glennis" after Yeager's wife. The date was October 14, 1947.

Yeager, nursing some bruised ribs, closed the door to the tiny X-1 and buckled in for a bumpy ride with destiny.

At 45,000 feet (13,700 m) the X-1 was released from the underbelly of its mother plane, a Boeing B-29, and it began to freefall. After a few seconds Yeager pushed the button that fired its jet engines and the X-1 lurched and rocketed. Yeager and the X-1 howled a bit as if on the world's greatest amusement park ride. This particular ride had already claimed the lives of over a dozen other test pilots. Some considered it a suicide mission.

Soon he radioed his speed, "Point 6, point 7, point 8" referring to the tenths of percentages closer to the elusive Mach 1, which is 750 mph.

At "Point 9" Yeager radioed as the tiny plane buckled wildly. The entire plane shook so wildly that Yeager wondered if the screws would shake themselves loose and the plane would fall apart. The air was banging on all of the windows as the X-1 was knocking at the door to Mach 1. Then "Boom!" the world's first sonic boom, as Yeager and the X-1 broke through the very racket they created.

The sound barrier was considered by many as an absolute, a limit that could not be broken. Yeager showed this to be a myth. He showed that it was an envelope to be pushed. Others would push the mach meter further with Scott Crossfield breaking Mach 2 in 1953. Fifty four years later, on October 6, 2007, the speed car Bluebird would break the sound barrier on land. In truth, the car resembles a jet rocket engine turned sideways, but I suppose since it has wheels it can roughly be counted as land based. Bigger, faster and farther, the race continues, with no end in sight. But for October 14, 1947, Yeager was the ace of aces, and it was a great moment of science.

67. The Dark Side
Luna 3 Photographs the Back of the Moon
October 6, 1959

The moon is in a rather unique dance with the earth. Its orbit around the earth and the time it takes to spin once on its axis, a moon day, is nearly equal, 27 1/3 earth days. Because of this, a strange thing happens. We only see one side of the moon. The backside, or dark side as its called, always faces away from the earth. If one watches the earth from the moon one would see North and South America for a while, then several hours later the Pacific Ocean scrolls by, and finally Europe and Africa. But this is not true of the moon. The earth only sees one side of the sphere, always the same side.

Thus it was with great anticipation on October 6, 1959 when Luna-3, an unmanned Russian spacecraft would be the first craft to circle around the moon and take pictures of its mysterious surface. The 778.5-kg (612-lb.) cylinder craft snapped 29 photographs but the quality of the pictures were only fair to inadequate. Were there any surprises? We were shocked at the number of craters as compared to our familiar side; nearly half a million more. Always facing out to the solar system, the far side of the moon was victim to many more collisions with wayward space debris and asteroids and it had the scars to prove it.

68. A Dog's Life
November 3, 1957

The first living creature launched into space was a Russian dog named Laika. It is unclear whether Laika actually volunteered for the job, but nonetheless he was strapped into a tiny capsule on November 3, 1957 and launched.

The Soviets held a news conference to announce to the world their achievement. This success they felt demonstrated the superiority of the Russian space program, its society and its communist government. Laika may not agree. When asked when the triumphant Laika would return to earth, the scientists look bewildered. This was only a one way trip! There was never any plan to return Laika to earth. He died several hours later still in orbit, so much for a comrade's best friend and animal rights, an important cause for many.

69. First Man in Space
Yuri Gagarin of the Soviet Union – April 12, 1961

The United States and the Soviets were in a race to put the first man in space. That honor goes to a Russian Cosmonaut by the name of Yuri Gagarin on April 12, 1961. It would only be two weeks before the United States had their own man in space, Alan Shepard who was launched on May 5, 1961 aboard Freedom 7. But April 12th belonged to the Russians, they had put the first man in space, an embarrassment that made America just a little bit red in the face over another Soviet first.

In an earlier Soviet flight, two dogs Belka and Strelkha were successfully launched but today the stakes were much higher. The courageous 27 year old flashed a broad smile as he stepped into the miniscule Vostok 1 capsule designed by Sergei Korolov, father of the Soviet space program. Yuri was seated horizontally, his back level to the ground, so that his heart would not have to pump upwards against the 9 g's, or nine times the gravity forces he was sure to encounter.

"Poyekhali" – "Here we go!" he said pumping a thumbs-up sign. This was to be an autonomous flight, controlled by the ground crew at Star City, 20 miles northeast of Moscow. The control stick had in fact been locked to ensure no pilot intervention. Most gave his chances of survival at 50%. Two press releases were prepared in advance, one for success and one for failure. He would be given an in-flight promotion to major, to help curtail the grief should this endeavor turn badly.

At 9:07 the mammoth rocket ignited sending its fiery breath shaking the Soviet earth. Reluctantly, slowly the Vostock 1 inched its way skyward shaking and roaring. Gagarin's face flattened a bit and he struggled to breathe as the weight of the added gravity pressed against his chest. Slowly at 9:18 the foot of planet earth lifted off of Gagarin and the capsule achieved orbit. Khrushchev later remarked Gagarin "flew into space, but he didn't see God up there."

Gagarin drifted over the Atlantic, and a sleeping America. At 10:15 Vostock 1 was over the Pacific and the reentry phase began. The ground control signaled the retro-rockets to separate out the service module from the capsule but something went wrong. The two pieces of the craft were still tethered by a set of wires connecting them. In a deadly tug of war the two crafts began to spin and rock spinning Gagarin helplessly inside. If the service module did not shake loose Gagarin would die. Chief designer Sergei Korlov could only pace as the troubled craft entered into the ionosphere where radio contact is lost. After 10 long minutes of silence Gagarin radioed that separation had been achieved, but the wires had melted in the heat of reentry. At 10:43, the pilot ejected as planned from the capsule and both drifted slowly to the ground thanks to their separate parachutes. They landed in Serbia after a 1 hour and 48 minute flight. No mention was made of Gagarin's near-death experience for 30 years and the flight was hailed around the world as a complete success. Gagarin praised it as a vindication of the Soviet political system which he dubbed the "organizer of all our victories." Gagarin died just a few years later on a routine flight on March 27, 1968.

70. The Engine
Nikolas Otto Invents the Internal Combustion Engine 1876

It has been long thought that Leonardo da Vinci had a working helicopter design as early as 1486. All that it lacked was an engine. In the center of the da Vinci's helicopter was a human being, feverishly pedaling a pulley system that would rotate the blades overhead. He could not hope to ever generate the foot speed required for actual flight.

The engine Leonardo and others would need was not invented until 1876 by a German inventor Nikolas Otto. In fact, gasoline itself was not discovered until 1861. The automobile was in its infancy. Early prototypes tried to parallel the success of the steam powered trains, and ships, but such a power source was cumbersome. Finally Frenchman Etienne Lenoir broke through with a two stroke automobile engine that could move at a modest 4 mph, or walking speed.

"Yes, gas, that is the answer," Nikolas Otto (1832-1891) mumbled as he read of Lenoir's two stroke engine. "Steam power was too cumbersome, better left for the big locomotives," he said with his strong German accent. He lifted up his well worn shoes atop his sample boxes of coffee, rice and sugar, which provided the bread and butter of his existence.

The Ettiene engine was not only underpowered but also irregular running, providing only jerking propulsion.

"No, someone can do better than that" replied Eugene Langen, a German industrialist who would soon be Otto's business partner. That someone would be Nikolas Otto.

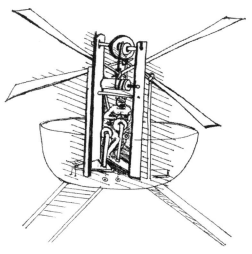

Leonardo da Vinci's helicopter lacked only an engine

As Otto looked out the train window he could see the smoke from a chimney and that was his first inspiration. "It must work like the hearth of a fireplace, where all the materials needed for combustion in the proper percentages are present." He envisioned a four-step engine cycle. First he would bring in proper proportions of air and gas into a central chamber. A second critical design point was missed by Lenoir, "compression." The mixture must be squeezed in order to generate a significant explosion. Third, the explosion energy would be transferred through the crankshaft to the wheels. In the final step, the exhaust would be forced out.

By the end of 1876, Otto had finished the engine. It became the envy of the world and the standard for 100 years. Although many modifications have occurred, Otto's 4 stroke basic design is used today anytime a need for quick compact power is required as in cars, lawnmowers, motorboats, farm equipment, manufacturing, snow blowers or generators. His first prototype produced a respectable 3 horsepower. It was equal to the propulsion provided by 3 horses. By the end of his life this output would increase to an amazing 100 HP, unheard of in its day. Over 30,000 Otto engines would be sold.

It's hard to imagine our modern world without the internal combustion engine. For one, the Wright Brother's flight of 1903 could not have been possible. Nikolas Otto did not invent the automobile, or many other advances of the industrial revolution, but he did contribute one of its critical organs, the heart if you will, that opened the valves of the modern world.

71. Archimedes' Ship
212 BC

In 208 B.C. Greek mathematician Archimedes watched as King Hiero's soldiers of Syracuse tried to launch a large war ship from the shore into the water. Dozens of strong-armed soldiers grunted and moaned but the enormous ship refused to budge.

"I think I can help," Archimedes offered. Archimedes was known far and wide for his genius, especially in mathematics.

"I think I can solve your dilemma with a series of levers and pulleys if you will allow me to try," he told the king.

"Give me two weeks, and all the rope and wheels I require, and I will let you personally launch your ship with one hand."

King Hiero laughed. He may not have understood exactly the scientific principles behind forces and pulleys, but Archimedes promise had peaked his curiosity.

"OK, Archimedes," Hiero said with a smile "you'll have your rope and all the luck one can conjure." Archimedes bowed his head meekly and departed.

"He would have enough rope to hang himself," Hiero laughed with his cohorts. That's the trouble with these academics, no real world experience."

Two weeks passed as large quantities of rope, wheels and timber made its way to the beachhead. Finally, Archimedes sent word to Hiero that the boat was ready for launch. Upon arriving at the beach, Hiero was immediately struck by the labyrinth of pulleys, ropes and timbers that stretched to the bow of the ship. If nothing else, this certainly was impressive.

"Okay Archimedes, you cost me enough rope, let's launch this ship."

"Not just yet King" Archimedes asked, "Can we first put all the cargo, equipment and men back on board?"

With a wave of his hand Hiero sent his soldiers packing. With a little bit of chaos all was ready. "Okay Archimedes, the ship is yours."

"No Sire, the ship is yours," Archimedes said as he handed Hiero the handle of the end lever. Hiero stared at it a moment and then with one hand pushed down on the first lever and the enormous pulley system groaned. The ship then lifted inches off the ground, dangled, and then came gently back down afloat in the water. The boat had been launched. A collective gasp was heard as others cheered.

Archimedes explained that pulleys are like levers that can multiply force. He even boasted that if he had a solid place to stand and a lever and a fulcrum, he could move the earth, "Do you see?" King Hiero did not see. But he did know that Archimedes was a genius of stupendous stature, perhaps the greatest mind of the ancient world. He pulled the lever just one more time and wondered "Just how does he do that?" But Archimedes wasn't listening. He had turned his head to watch a few birds fly effortlessly around the beach. He was wondering how the birds did that.

72. Hospital Standards
Joseph Lister Pioneers Antiseptic Surgery
1867

There was a killer that stalked the hospitals of the 1800's. It was called septicemia, gangrene, or just plain old "Hospital's Disease." Too often doctors would record "Operation a success, patient died of infection." Infection was so prevalent that hospitals were seen as places to die not recover. Pregnant mothers, amputees, abscessed teeth, and compound fractures were amongst the worst hit. At some hospitals, death by infection was as high as 60%.

One doctor, Englishman Joseph Lister vowed not to just stand by. "Microbes were the enemy!" declared Louis Pasteur in 1865. Lister insisted that doctors wear white robes to show the dirt. He asked that they scrub often, especially between patients. He asked that instruments be sterilized and carbolic acid be sprayed during operations to kill lurking germs. He had heard that carbolic acid was used to treat raw sewerage and thought it must work to fight bacteria. Under Lister's guidelines he cut the mortality rate in one hospital in Glasgow from 60% to an incredible 4%. Much of Lister's antiseptic practices are in use in today's hospitals. The oral antiseptic Listerine is named after Lister.

73. Voyager
First Man-Made Objects
Leave Solar System
1990

As the story goes, a man stranded on a deserted island longs to contact the outside world slips a note into an empty bottle and casts it into the sea. He knows well that his chances are slim that anyone will ever retrieve his message but still there is that tantalizing possibility that some favorable tides will bring him help.

In the 1970's, NASA sent its own message in a bottle. Four unmanned spacecraft, Pioneer 10, Pioneer 11, Voyager 1 and Voyager 2, completed their solar system missions and then set sail for points beyond Pluto. They became the first man-made object ever to leave the solar system. By 1990 all had exited beyond Pluto.

In the future, giant spaceships will billow their sails with solar winds bound for worlds unknown

In fact, the message in a bottle had a much better chance of being discovered than Voyager since it needed to go just a few thousand miles. The oceans of space are much vaster in comparison. If one were to shrink our solar system down to the size of a Frisbee, with the outer edge representing the orbit of Neptune, the next Frisbee of star, Alpha Centauri would be located some 45 miles away. The next Frisbee would be over 100 miles away. Even at 10 miles per second, our crafts may have to travel 80 thousand years before reaching Alpha Centauri, 4.3 light years away.

Like the bottle, our crafts too carried messages to outside worlds should any one find them. On the Pioneer 10 and Pioneer 11 are plaques 9 inches by 6 inches with the figures of two humans male and female. It also shows our relative positions in the solar system and some signature binary stars. On board Voyager 1 and 2, is a 12 inch gold record containing a myriad of sounds found on earth. There are 55 different people saying "hello" in different languages, a baby crying, a dog barking, birds, and thunder. There are also selections from Chuck Berry, Louis Armstrong, Bach and Beethoven…familiar sounds as it takes its long day's journey into night.

74. Frog Power
The Invention of the Battery - 1799

In 1780, Italian Luigi Galvani (1735-1798) settled into his routine dissection of a frog. Just as he began cutting, the kicking of the frog's leg startled him. "Oh, my God," he said taking a half-step back. He wasn't sure if his frog had died. A closer inspection confirmed that the frog had indeed died. What caused the frog's leg to convulse? After a bit of trial and error, he was able to repeat this phenomenon. "The frog must have some 'natural' electricity" he thought, a sort of "frog power."

A curious Alessandro Volta (1745-1827) of the University of Padua repeated the experiments but found that the electricity was not resident in the frog but was caused by the presence of two dissimilar metals. The frog had been pinned down with brass clips and the scalpel was iron. The electrons would leave the surface of one metal to try and reach the other metal causing a flow of current which convulsed the frog's leg through its nerves. In fact, he didn't need the majority of the frog at all, just its legs. He proved this graphically by removing most of the frog leaving just its legs and they still kicked.

Volta took his research one step farther in 1799 when he invented the voltaic cell, otherwise known as the world's first battery. It was a series of small discs with a hole in them piled one on top of the other. Each alternating disc was silver then zinc and then silver again. They were separated by wet cardboard and connected in a series which multiplied their voltage.

Alessandro Volta's battery was a major breakthrough in the study of electricity. Before him electricity could only be studied haphazardly using glass Leyden jars. It stored electrical energy through friction but it was far too unpredictable. It needed to be discharged fully all at once. With Volta's battery, we had a method to generate consistent and measured current. He helped to jump start the serious study of electricity.

75. Paperclip
May 2, 1945

A nervous Magnus Von Braun pedaled his bicycle down the mountainside. He was fleeing Nazi Germany looking for the Americans. General Eisenhower was looking for him, or more specifically, his brother Werner von Braun, who headed the Nazi V-2 rocket program.

"Wait don't shoot!" a nervous Magnus shouted. "My brother, Werner and his V-2 rocket team want to surrender to Ike," he said. Germany was collapsing. Adolf Hitler had committed suicide two days earlier in a bunker in Berlin. Hitler had left standing orders for the SS to execute Von Braun and his entire rocket team in the event of his death. In a last ditch effort Magnus was here to arrange their surrender.

This surrender began "Project Paperclip," an initiative by President Truman to bring Germany's V-2 rocket team to the United States. On this day of May 2, 1945 the United States acquired a treasure trove of rocket experts, over 100 scientists, in fact almost the entire Peenemunde V-2 rocket team. Dr. Werner Von Braun, the genius who headed up the V-2 team,

arrived sporting a long arm cast from an auto accident earlier in March.

Along with all of the top scientists, Project Paperclip would gather over 14 tons of priceless engineering documents and enough spare parts to build over 100 V-2 rockets. Werner Von Braun went on to be the father of the American Space Program and the first head of National

Aeronautics Spaces Agency (NASA). It is estimated that the acquisition of these scientists pushed the United States space program ahead twenty years in the development of rockets. It was truly a great moment of science for Americans. The landing on the moon did not start with President Kennedy's speech in 1961; it began with a bike ride down a hill on the Austrian border in Germany in 1945.

76. Tesla's Death Ray
Tesla Claims to have Invented a Death Ray
1938

"I have built a Death Ray," began the gaunt 78 year old Nikolai Tesla in 1938. The reporter from the New York Times didn't quite believe him but he knew a good headline when he heard one, Tesla was no ordinary septuagenarian. He was one of the most recognizable men in the world, holder of over 800 patents, and a primary architect of our AC power world.

"It can destroy 10,000 enemy planes at once for up to a distance of 250 miles. After that the curvature of the earth would limit its effectiveness," Tesla explained. His Serbian accent was still intact after fifty plus years of living in the United Sates. He arrived in New York City in 1884, penniless and alone, only to make a fortune and then lose it again. While he spoke English today, Tesla could have called upon any one of several languages in which he was fluent.

The "Death Ray," he explained "is based on the principle of natural frequency. Everything has a fundamental frequency. One has simply to find this fundamental frequency and exploit it." Tesla moved his arms to show how two natural frequencies must be brought into harmony for the weapon to work.

"There is no defense against such a weapon," he continued. Tesla would send a concentrated particle-beam through the air by means of a new 'teleforce,' a force he says that was hitherto unknown to man and is based on an entirely new principle of physics. Tesla would not elaborate. Most believed he had stumbled upon the idea of laser weapons. The "laser" travelling at the speed of light would strike its target nearly instantaneously blowing apart the plane and its cargo.

"Isn't it true, Mr. Tesla that much of this 'Death Ray' work was done in Colorado Springs in 1889?" probed the reporter. Tesla sat purposely closed lipped but the twinkle in his eye said that it was so.

"No comment," Tesla responded.

Tesla offered to sell his Death Ray for $20 million exclusively to the Allies, and not Germany. No country, including the United States, England, Canada, France or the Soviet Union responded favorably to this offer.

It is difficult to say why no one would buy it, except perhaps that no one quite believed him. Obviously, such a weapon would have been of unfathomable value during World War II. But Tesla was not an easy man to take at his word. He was prone to making grandiose statements and would not demonstrate his weapon prior to its purchase. He provided no working prototype, insisted on complete autonomy and secrecy, and he was asking for very large sums of money.

Did Tesla's 'Death Ray' really work? We will never know. Tesla died just a few years after this interview in 1943 and the weapon was never seen or demonstrated. There is however, a tantalizing ending to this "Death Ray" story. Tesla died on January 7, 1943 at the age of 87. A few days before he died, he ostensibly showed a strange box to a pizza delivery boy at his hotel room in New York City.

Tesla told the boy that the box contained an incredible new weapon of immense power. An overwhelmed delivery boy nodded quickly and then wisely made haste to leave.

It is curious however that in March 23, 1983 President Reagan proposed an ambitious project known as Strategic Defense Initiatives (SDI) otherwise known as "Star Wars." It featured a space-based anti-missile system that would shoot ballistics with laser beams. Did they borrow from Nikolai Tesla? Upon his death, the United States Government, under the direction of J. Edgar Hoover and the FBI, seized all of his papers and belongings. Copies of his 'Death Ray' papers were sent to Wright's Field in Dayton, Ohio. They began a super secret project known as NIK (short for Nikolai Tesla) to recreate his 'Death Ray'. What was the result? No one is saying. Whatever Tesla's secret was, he took it with him. Perhaps even he couldn't build it.

77. Discovery of Cells
1665

In 1665 Englishman Robert Hooke (1635-1703) took out his one of a kind compound microscope. He had built it himself and it was amongst the finest in the world. He took out his penknife and shaved a slice of cork a couple of layers thick. He carefully positioned it beneath his apparatus and turned the lenses slightly to bring his captive into focus.

**Robert Hooke
(1635-1703)**

"Honeycomb!" he exclaimed referring to the cell like structure that bees make to store their honey and brood. "It is lots and lots of individual cells close together and empty."

The cork cells Hooke was observing had been dead for a long time and thus had no nucleus. It was later found that all living organisms, plants and animals, are made up of cells. It was Robert Hooke who first provided these cellular stepping stones to today's modern biology and medicine. Hooke carefully sketched his subject, which would later be in his book 1665 book *Micrographia*. Hooke was the first one to discover cells and coin the term to describe the basic building blocks of life.

78. Fission Vision
Lise Meitner Conceptualizes Nuclear Fission
December 24, 1938

In 1937 Enrico Fermi and others began bombarding the nucleus of the heavier atoms, namely uranium, with neutrons. They had hoped that some of the neutron bullets would stick to the nucleus of the atom increasing its atomic number from 92 to 93 thus creating a new artificial element.

A review of the experimental data by German Otto Hahn and Fritz Strassman showed some puzzling results. Instead of getting Neptunium which they had expected they got Barium which has an atomic number of 56. "How could the atomic number have gone down," they puzzled "and so dramatically?"

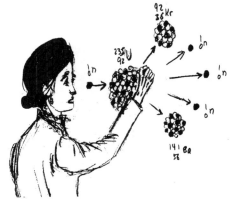

**Lise Meitner
1878-1968**

Otto sought the help of Lise Meiitner his intellectual partner of 30 years now in exile, Lise Meitner was Jewish. When Hitler came to power in 1933 she was forced to flee Germany and take up residence in Stockholm.

On December 24, 1938 Lise Meitner sat spellbound in the lobby of her hotel after reading a letter from Otto Hans.

"Ah, the atom must have split!" she said with a wave of her hands.

"No, that would be impossible, atoms don't split" replied her nephew Hans Frisch. It was a truly radical idea, one that had not occurred to Otto.

"No, think about it Hans. Barium has 56 protons and Krypton, the other small piece, has 36 protons. They add up to 92, the exact atomic number of Uranium!"

Lise Meitner and her nephew Hans Frisch would publish their results in 1939. Hahn and Strassman would confirm it experimentally that same year. They received the Nobel Prize for it in 1944.

If Robert Oppenheimer is considered father of the atomic bomb, then Lise Meitner was its mother. It is said of Lise that she left Germany with the bomb in her purse. Einstein called her "Germany's own Marie Curie."

79. Germ Warfare
Louis Pasteur Finds a Vaccine for Rabies
July 6, 1885

A very distraught Mrs. Meister climbed the few the stone steps leading to Louis Pasteur's laboratory on Rue d'Ulm. The date was July 6, 1885. She had traveled with her sick son all the way from Alsace in western France to Paris to see the great Louis Pasteur. In tow was her nine-year old son Joseph, who had been bitten by a rabid dog two days earlier.

She knocked on the door and a well kept Louis Pasteur ushered them in. Pasteur walked with a limp as a stroke in his fifties left him partially paralyzed on his left side. She pulled the handkerchief from her mouth and face from which she was wiping copious tears.

"Please sir, save my son!" she implored, "He's been bitten by a rabid dog."

Pasteur grimaced in sympathy as he examined Joseph's nasty wounds fourteen of them in all. Joseph could barely walk. Rabies is one of the most horrible diseases. It is also called hydrophobia because patients convulse as water is brought to their lips although they are extremely thirsty. It is a

disease of the central nervous system that gives the feeling of strangulation, paralysis, seizure, and finally death. It enters the human body as a result of bites mostly from dogs, wolves, skunks, raccoons, bats, and cats. In the 1880's it was a death sentence. No one had ever survived.

Pasteur was in a great moral dilemma. He was not a true doctor but a chemist by training. Should something happen to the boy during his treatment he could lose everything. Mrs. Meister begged him, pleaded with him, as only a mother who loves her son can. There was a rumor that Pasteur had a new cure for rabies by enlisting the body's immune system. Edward Jenner first pioneered this strategy with small pox in 1796. Pasteur consulted with Dr. Vulpian and Dr. Grancher who agreed to oversee his treatment.

"If you don't treat the boy he will surely die" argued Dr. Vulpian. Pasteur's humanity carried the day and he agreed to treat the boy. He had not given the vaccine to humans, only dogs and rabbits. But this was a human being, a boy named Joseph. The stakes could not be higher.

At 5 PM on July 6, 1885 Louis carefully prepared a set of rabid syringes. He held them upside down and flicked each with his index finger to make sure there were no bubbles. He would need to inoculate for fourteen days straight in order for this to work. Starting with the first, in each syringe was a weakened rabies virus so that the body could develop immunity to the virus. The next day's inoculation would have a stronger virus, and then stronger still on the next day. Why the fourteen days? Rabies needed fourteen days to travel from the site of the bite to the central nervous system in the body.

On the first night he inoculated Joseph Meister and sent him home. Pasteur got very little sleep that night awaiting the fate of his patient. Louis Pasteur, a religious man, spent most of the night praying.

The next morning Joseph Meister showed no signs of deterioration. Pasteur continued to give Joseph injections for fourteen days. The boy's health continued in tact. An elated mother and child returned to Alsace and never exhibited a trace of the dreaded rabies. Louis Pasteur made Joseph promise to write him regularly so that he could keep track of his health.

When word of this healing became known throngs of people flocked to Paris seeking a cure. Boxes and boxes of correspondence begged Pasteur to visit their town and cure their sick people of rabies. Once, nineteen peasants from Russia came to his door, all of them bitten by a ravenous wolf. It had been nineteen days since they had been bitten, plenty of time for the disease to have made good headway towards the spinal column and brain. Miraculously, sixteen of the nineteen were cured. A grateful Tsar sent one hundred thousand francs with which Pasteur built one of the great medical institutes of the world in Paris, the *Institute Pasteur*.

Pasteur along with Jenner had finally struck a blow against the microorganisms that plagued mankind. His was not the first vaccine but certainly one of the more important. Other vaccines would follow including diphtheria, tetanus, measles, mumps, rubella, meningitis, and the dreaded polio disease to name just a few. For the first time in history, man had a weapon against rabies. There's an old saying that a cure can be found by taking a little hair from the dog that bit you. It wasn't invented by Pasteur, nor is it a serious statement of theory, but it certainly applies.

80. Quarks
One Fish, Two Fish, Red Quark, Blue Quark - 1964
1964

Protons, neutrons, and electrons; we thought we had gotten to the bottom of things, the ultimate building blocks of the universe. Then cracks in the atomic foundation began to emerge. Marie Curie and others discovered that all of a sudden, a perfectly good atom, such as uranium or radium would spit out protons and neutrons in what is known as

radioactivity. Huge particle accelerators were creating fractions of atoms. It was clear that our pristine model of the atom had little cooperation from nature.

In 1963 Murray Gell-Mann proposed that protons, neutrons and electrons were not the most basic particles. They in turn were all made of a smaller set of particles called "quarks." No one quite believed Gell-Mann until in 1968 when Jerome Friedman, Henry Kendall and Richard Taylor actually found one at the Stanford Linear Accelerator Center.

What is alarming is that the number of quarks found is growing. We now have found a "charmed" quark, and a "strange" quark, a "top" and a "bottom" quark. What is clear is that particle physics right now is a mess waiting for a theoretical synthesis. Hope is riding on a new Super Collider to be built in Texas with a 54 mile race track for subatomic particles to interact. As of 1993 this Super Collider's funding of $12 billion has been tabled. Until it is refunded, we will just have to get along in an imperfect world and put up with each other's quarks.

81. Archimedes Death Ray
212 BC

Sixty-four Roman ships were poised under the direction of General Marcellus for an all out assault on the port city of Syracuse, Sicily. The year was 212 BC. An eerie silence was broken by the creaking of the wooden boards.

"There is a madman on that island" General Marcellus grumbled. That "madman" was Archimedes (287-212 BC), a genius of historic proportions. He had already shown that he had more than a few tricks up his sleeve. In the last assault large grappling hooks came out from the city walls and lifted and shook his ships mercilessly in the air. Then suddenly,

"Fire!" was screamed from one of the ships. Marcellus turned to see men beating out flames with their robes trying to snuff the flames.

"Fire!" a second ship screamed as one of its oil drums burst into flames. Suddenly a third ship was engulfed.

Up on the city walls Marcellus could see old Archimedes turning what appeared to be a second sun. He and several others were focusing the sun on their curved shields used as mirrors setting the Roman ships ablaze.

Chaos was everywhere and Marcellus sounded the retreat. Eventually the Roman's would prevail, but for this day Archimedes held them at bay.

This story comes to us through historian Plutarch who was writing about the life of Marcellus. Is it possibly true? In a 2005, a television show called "The Myth Busters" (episode 16) , scientists tried to recreate Archimedes famous Death Ray and failed. But on September 30, 2009 scientists from Massachusetts Institute of Technology (MIT) were successful in igniting a replica of the Roman ships under very controlled conditions. Did it happen? We can only guess, but it's too great of a triumph for science not to reflect on it.

82. Deep Blue
A Computer Defeats the World Chess Champion
May 11, 1997

On May 11, 1997 World Chess Champion Garry Kasparov stared through the chess board. He did not like what he saw. He peaked over his folded hands to see if his opponent noticed his weakness but there was nobody there. In fact, his opponent was a computer called "Deep Blue" developed by IBM. "Big Blue" is IBM's corporate nickname. "Deep Blue" was a combination of "deep thoughts" and "Big Blue."

They had played five earlier matches. Kasparov had won the first match, and Deep Blue the second. Three ties in a row had brought them to this decisive 6th game in New York City. A throng of chess enthusiasts had gathered in Suite 423 of the Equitable Center to watch every move.

Kasprov had tried a risky strategy, swapping a knight for position. He reasoned that the computer would not enjoy playing at a material disadvantage. But now just eighteen moves into the game Kasparov had lost his queen and then resigned.

A frustrated Kasparov finally pushed back on the table and walked out of the room. Big Blue's human proxy C. J. Tan smiled as he and his team accepted victory.

Was this a great moment of science? I guess that depends on which side of the sixty-four squares you're sitting on. A machine had beaten humanity in its strong suit, creative intelligence and it is hard not to take such a loss personally. Was Deep Blue able to savor its victory? Or was it no more conscious than a television set is conscious of the pictures it's flashing? It's a deep question. For today this smarting defeat for humanity is best forgotten, buried somewhere in memory.

83. Planet Vulcan?
Circa 1800

At the turn of the 20th century astronomers and amateur astronomers alike were in a frantic search for a mythical planet called Vulcan. Vulcan was the name of the Greek god of fire, and so it seemed a fitting name, for this alleged 10th planet since it was purported to have an orbit very close to the sun, making it very hot.

The search began when the orbit of Mercury was off by the slightest of degrees. It wasn't much to speak of really, just one one-hundredth of a degree (43 seconds of an arc) every one hundred years, but it was there. It was such a small precession that it would take over a million years before Mercury would trace out its characteristic flower pattern in its yearly orbit around the sun. Scientists knew that such an anomaly must have a cause; perhaps the pull of an unseen planet and thus the search for planet Vulcan began.

The vigil for planet Vulcan lasted for

Scientist reasoned that Mercury's precession was caused by a 10th planet

121

about one hundred years. No one ever found it, although many claimed erroneously to have sighted it. The search for Vulcan ended in 1916 with the work of Albert Einstein on General Relativity. Einstein's work replaced Newtonian gravity with a curved four dimensional space-time. The two systems predicted nearly identical orbits for all the planets with the exception of those very close to the sun. Einstein's four-dimensional model of space-time predicted exactly the precession of Mercury. There was no longer a need for a planet Vulcan.

Still, some dreams die hard. Some people continued to search for Vulcan and its legend has gained almost mythical status. In television culture, one of the main characters of Star Trek, Mr. Spock, was purportedly from the planet Vulcan. If on some sunny day you happen to notice an extra tiny black spot crossing the surface of the sun, or if your neighbor's ears appear to be a little more pointed then is customary, perhaps you are seeing evidence of planet Vulcan, a myth that continues to "Live long and prosper."

84. The Turtle
Bushnell Builds the First Submarine
Sept 7, 1776

"It was a crazy plan," admitted General Putnam on the night of Sept 7, 1776. "To sink a British warship using a submerged barrel," he shook his head. The oversized barrel built by Connecticut native David Bushnell would be the first submarine. Tonight's target was the British 64 gun frigate, the HMS Beagle, anchored in New York's harbor. The Turtle quickly traversed the one mile distance that separated him from his prey anchored in New York's harbor.

It had been nicknamed "the Turtle" because it resembled a turtle's back when it was in the water. It was equipped with a corkscrew on its top and in front with which to drill a hole into an unsuspecting ship. It also carried a 150 pound powder keg. Constructed almost entirely of wood, it was four feet wide and seven feet tall. It could comfortably fit only one person at a time. It had a hand crank which propelled the tiny craft forward.

Sergeant Ezra Lee quickly crossed the water and submerged under the great ship. He was not able to drill through the thick skin of the enormous boat but he did make enough of a racket to alarm the crew. Lee frantically pedaled as the British gave chase. The British were closing in on him when in desperation; Lee released the 150 lbs of explosives. A huge plummet of water went skyward drenching the entire area and providing enough of a diversion for Lee to escape. A startled British navy retreated their entire fleet of 200 ships several miles away until they could figure out exactly what just happened. When Washington heard of Bushnell's attempt it was said to be one of the few times that he ever laughed during the war.

The Turtle tries to sink British frigate HMS Beagle

Today the submarine, pioneered by Bushnell, is an essential part of any modern day navy. Its stealth and ability to move undetected through the water remains one of its key attributes. It provides an important technological edge to help keep one's enemies off balance and over a barrel.

85. Charles Babbage Designs the First Computer
1821

Charles Babbage (1791-1871)

Charles Babbage was a numbers guy. He counted everything from the number of broken windows per week to the number of loaves of bread consumed in the marketplace. By 1821, he had one of the largest private collections of printed tables of numeric values in London. When he found error after error in his tables he was beside himself. A typical volume would average about 100 errors. Babbage blamed the computers. Not the 1970's silicon variety of computers, but Victorian era computers, people paid to compute by hand table values. He resolved to create a machine to do the computing, or the world's first computer.

It would take Babbage over 20 years to come up with his Difference Engine. It was to be about 8 feet high by 7 feet long by 3 feet deep and weigh as much as 12 cars. . It would add, subtract, multiply and divide. Unfortunately, only a small portion of it was ever built. His design had over 25,000 intricate parts including counters, gears, wheels and pulleys. Many of them could not be produced to tolerances that his machine required and he did not have the money to build all of it. He had a portion of it built, but then quarreled with the machine maker who then quit.

Undaunted Babbage moved on to an even more ambitious machine called the Analytical Engine. If the first required a large room, this new machine would require several rooms and a steam engine to operate it. It was never built. It contained several features seen in the first modern computers. Like Hollerith's computer, Babbages' Analytical Machine used punched cards as input, and could be programmed.

He died in 1871, a small cog in the great machine we call humanity. He was passionate about machines, numbers, and accuracy. Due to the lack of tolerances of the age he could not make the Difference he had hoped for humanity.

86. The Speed of Light
Roemer Measures the Immeasurable
1674

Galileo had a notion that the speed of light was finite. He thought it was very fast but finite none-the-less. Is the speed of light infinite or finite? This was the great debate. Did it travel from point A to point B at a certain velocity or was it instantaneous? Aristotle felt that its speed was infinite. Galileo felt that it was not and set up to prove it in 1667.

Galileo set up two lanterns one mile apart between to distant hills outside of Padua, Italy. When Galileo flashed the light from his lantern, his assistant one mile away would send back a corresponding flash of a lantern. After many tries and much practice, Galileo could find no appreciable lag time other than human reaction

**Ole Roemer
(1644-1710)**

speed. He concluded that while he still believed light had a finite speed, it was much too fast to be measured over a distance of one mile. In fact such a distance would have been covered in less than 5 millionths of a second.

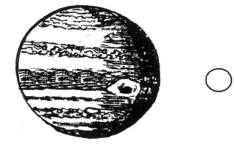

Following Galileo's lead, in 1675 Danish astronomer Ole Roemer (1644-1710) had a plan. Galileo had the right idea, but the distance would need to be much greater and the lanterns would need to be brighter.

Roemer would ingeniously use the distance to Jupiter's moon Io to measure the speed of light. Io orbited Jupiter every 42.5 hours. From earth, Io should disappear from view behind Jupiter and reappear on the other side exactly 42.5 hours later. However, every night it was off by fourteen seconds. For six months, as the earth approached Jupiter, Io would reappear 14 seconds earlier than expected. Over the next 6-months, as the earth was moving away from Jupiter, Io's reappearance would be delayed fourteen seconds. He knew that Io's very reliable orbit had not changed. The additional 14 second delay was the time it took light to travel the extra distance to the earth. The earth had moved away during it annual sojourn around the sun. Since he knew that extra distance and the extra time he could now calculate the speed of light.

Roemer was able to calculate the speed of light to within 26% of its actual value. He measured it to be 2.2×10^8 meters per second. Its actual value was 3×10^8 meters per second, or 186,000 miles per hour. In fairness to Roemer, the accepted distance from Jupiter to earth was in error. If he had a more accurate number to begin with he would have been much closer.

Eventually, in the 1880's American Albert Michelson and Edward Morley would provide science with the first accurate measurement of light. They received the Nobel Prize in 1906 for their work. But Roemer's experiment was one of the most resourceful ideas since Eratosthenes measured the size of the earth with a few sticks and some shadows. Today, the speed of light is one of the most important constants in science. Thanks to Galileo, Roemer and Michelson we at least know the speed of light, but its mysterious essence remains, a future task left for others to be found somewhere in the shadows.

87. DNA Fingerprinting
September 12, 1984

Connecting a suspect with a crime scene has always been critical in a police case. Law enforcement took a giant leap forward in 1856 when William Herschel discovered the uniqueness of fingerprints. Police were able to connect the perpetrator with a crime scene since no two people, not even twins, have the same patterns of tiny ridges on their fingertips. But clever crooks soon took to wearing gloves and fingerprints were not always present or obtainable at the crime scene.

On September 15, 1985, Alec Jeffreys of England discovered DNA fingerprinting.

The chances of a duplicate DNA match are about 1 in 100 billion.

Unlike fingerprints, DNA evidence is almost always present at a crime scene. It is in skin, hair, saliva and bodily fluids including blood and semen. A perpetrator almost always leaves some DNA at a crime scene. In the case of a murder, skin is often found under the fingernails of the victim as they tried to thwart their attacker.

Notorious Jack the Ripper may someday be identified by the saliva he left on the stamps

The very first criminal caught using DNA evidence was England's Colin Pitchfork in 1987. He was convicted of the rape and murder of two young girls, one in 1983 and one in 1986. He later confessed and as of 2007 is still behind bars. DNA testing has become the greatest law enforcement tool since the discovery of fingerprints. Some hope to identify the mysterious Jack the Ripper from the DNA found in the saliva he used to seal the envelopes that he left for police. Police are using this new tool to solve previously unsolved crimes, and extend the long arm of the law beyond ones fingertips.

88. The Hopeless Diamond
1966

Since the invention of radar during World War II, the search was on to build an "invisible" plane, that is, "invisible" to radar. The ability to elude such radar is known as "Stealth Technology."

In a twist of the Russian's stealing military secrets from the United States, the "Rosetta Stone" of stealth technology actually came from a Russian, Pyotr Ufimstev, in 1966. It turns out that the perfect shape for stealth technology is a pyramid. Incoming waves from a radar source can be deflected by pyramid shapes into what he called "re-entrant triangles" and are reflected internally into the wings of the craft and dissipated. Thus almost all crafts that are designed for stealth have this basic pyramid shape.

But designing a pyramid plane for radar evasion and getting it to fly are two different things. The United States turned to the crack engineers at Skunk Works who it's said could make a washing machine fly. They dubbed the project the "hopeless diamond." Bill Schroeder, a mathematician for Lockheed Martin, which also owns Skunk Works, developed a computer program that could exactly predict a plane's radar cross section (RCS). They successfully built the F-117A Stealth fighters and the B-2 Stealth bombers. In the 1980's both were used extensively during the Gulf Wars. They went virtually undetected to knock-out command and control targets in Kuwait and Iraq. In truth, the plane is not completely invisible, it has about the same cross sectional area as a small bird,

Pyramid Power
All stealth crafts have a pyramid shape that dissipates radar waves.

but it is still most impressive. The most telling testimonial to its invisibility to radar is the numerous bats that littered the runways after takeoff. Bats that fly with their own radar, known as echolocation, could not detect the planes and flew headlong into them.

Stealth technology was not only used for airplanes. The first stealth submarine, the Virginia, was commissioned for service in 2003. There is also a stealth ship Sea Shadow, which was the inspiration for the stealth ship in the James Bond film **Tomorrow Never Dies** (1997). What's next, a flying stealth washing machine?

89. What Happened to the Brontosaurus? 1975

Since its supposed discovery in 1876, the Brontosaurus had always been among the people's favorites. It's tough to say why, really. It could be that it was a plant eater, and thus much more approachable if such a thing is possible for a creature weighing the equivalent of twenty-five elephants. It could be that it had flat peg like teeth used for grinding vegetation was far less threatening than the razor sharp incisors of the Tyrannosaurus Rex. Perhaps one could imagine taking a ride on its mammoth back like some sort of prehistoric camel. Around 1975, all references to the Brontosaurus began to disappear from the reference books.

In 1876 American dinosaur hunter O. C. Marsh unearthed a large skeleton just outside of Como Bluffs, Wyoming. Unfortunately, his nearly complete skeleton was missing a head. A thorough search of the site came up empty until nearly half a mile away a dinosaur head was found. Since dinosaur bones weren't exactly as plentiful as chicken bones, Marsh concluded that this must the missing Brontosaurus head? It was not. In 1905 Marsh's carefully reconstructed the dinosaur bones, complete with the incorrect head, and dubbed it Brontosaurus. There it remained for the next one hundred years at the Carnegie Museum until 1975.

In 1975, two researchers, Jack Macintosh and David Berman, after twenty years of research, concluded that the Brontosaurus skeleton was a mistake. What was being represented as a Brontosaurus was actually a composite of two other well known dinosaurs, the head of a Camarasaurus and the body of an Apatosaurus. In fact, there was no such thing, nor had there ever been a Brontosaurus. An embarrassed scientific community quietly began removing all references to the Brontosaurus. A quick look up in the dictionary simply says "see Brachiosaurus."

Why is this a great moment of science? Sometimes a great moment is a quiet contribution of scholarship that overturns an entrenched thought. It shows that science is always subject to change and revision. Theories are simply that. As new evidence is presented science evolves. It is a living breathing changing entity, which is more than can ever be said of the Brontosaurus.

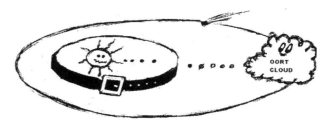

90. Halley's Comet 1705

Halley's comet appears once every 76 years

Although Halley's Comet is not the only comet in the solar system, it is the most famous and one of the few with a reasonable return time. By reasonable I mean within one human life time, which is 76 years. Sir Edmund Halley correctly concluded in 1704 that the 1683 comet was the same one noted by Kepler in 1607 and by Petrus Apianus in 1531.

It is even said to have influenced the Battle of Hastings in 1066. It was considered a bad omen for King Harold who in fact did die during the battle. Even famous American humorist Mark Twain commented on Halley. "I came in with Halley's Comet and I expect to go out with it" Twain was born in 1835 and died in 1910. Both of these years were appearance years also for Halley comet.

On February 7, 1999 NASA launched Project Stardust to fly through the tail of a comet and collect a sample of its tail. The word comet comes from the Greek meaning "hair of the head" referring to how the tail resembles long flowing

Mark Twain says "I came in with Halley's Comet in 1835 and I expect to go out with it (1910)".

hair. The comet selected was not Halley since it would not be back for another 63 years. Instead it chose the Wild-2 comet.

After a seven year successful mission, Stardust returned its sample canister to earth on January 15, 2006. It revealed comets to be made of the most ancient of celestial materials as old as the universe itself. Halley and others originate from a enormous cloud of dust beyond Pluto known as the Oort Cloud. They are still some of the most mysterious travelers of our solar system, and like Mark Twain, they have made the world a bit more interesting.

91. Space Elevator? Carbon Nanotubes - 1991

Putting an object in space is always difficult in terms of fuel spent, expense, predictability and danger. It takes approximately $50 for each penny one puts in orbit. Luckily, this may be changing with the discovery of carbon nanotubes, a substance 100 times stronger than steel.
Pioneered in 1991 by Sumio Iijima of Nippon Corporation and others, it has revolutionized the way we think of reaching space. One such idea is a space elevator. Instead of launching vehicles with an explosive burn, one might simply hit a button and up to the space station we go. A space cable will be tethered from a satellite in orbit and allowed to unfurl thousands of miles down to the earth's surface. Magnetic levitation trains would then shimmy up the cable like eager students up a gym class rope.

The problem is that the weight of such a cable 500 miles high or greater will collapse under its own weight. Imagine a column of Jello a two feet tall. Now imagine this same column of Jello a mile high! It would never stand up under its own weight. Building materials have always restricted the height that buildings can achieve. With steel we can now build skyscrapers much higher than concrete but steel also has a limit. Diamonds are our strongest material but they are brittle and costly. Diamonds however are nothing more than pure carbon. One can think of Carbon nanotubes as crushed diamonds rolled into thin layers and curled into a tube. It may be strong enough for such a space cable. The problem is that while we need a cable say a few thousand miles long, current technology has only allowed us to construct a carbon nanotube a few centimeters long.

NASA projects a possible completion date of a Space Elevator at the year 2100. NASA has also funded a challenge, the Beam Power Challenge.

In the Beam Power Challenge one needs to raise 25 kg up a tether at 1 meter per second for 50 m. The catch is that traditional sources of fuel or power cannot be used. This includes electricity, gasoline, oil, or natural gas. One must use energy sources readily available or easily obtained in space such as solar power or microwaves. Still the possibility is intriguing and the sky is the limit.

92. The Noble Prize - 1901

In 1888 Alfred Nobel opened up a French newspaper and read its shocking headline "Merchant of Death Dies." It was disturbing for two reasons. The first was that he was obviously still very much alive. They had mistaken his death with the recent death of his brother Ludwig. The second was that the paper referred to him as the "Merchant of Death."

"Is this my legacy?" He asked. "Had they completely misunderstood him?" Sure, he had invented dynamite and had made lots of money on it, but "Merchant of Death?" Had they missed the fact that his dynamite was also an invaluable tool to move mountains of earth and stubborn rock for railroads, canals and roads?

The damning headlines haunted him. He became severely depressed and he could not sleep. In a conversion worthy of Ebenezer Scrooge, Nobel vowed as Dicken's says to make "mankind his business." In his will, he established a series of Nobel Prizes to be given each year in his native Stockholm. He left about 94% of his net worth to fund these awards. The first Nobel Prizes were awarded in 1901, five years after Nobel's death in 1896. The establishment of the Nobel Prizes was a boon for scientists and a great moment of science.

**Alfred Nobel
(1833-1986)**

93. Einstein's Compass - 1884

Hermann and Pauline Einstein were quite worried about young Albert Einstein. The 5 year old barely spoke and when he did he would say the strangest things. He was told that he would soon have something new to play with. His parents were referring to the birth of his younger sister Maya. When Einstein was presented with his new sister he replied "Where are the wheels?"

One day when young Albert was sick, his father presented him with a toy compass. Hermann, an engineer in Munich, thought it might amuse him. Albert studied it intensely. Try as he might, and turn as he might, he could not get the needle to point anywhere but north. When Einstein learned that this was no parlor trick, that in fact it was the universe controlling the needle, Einstein was hooked. He said of that moment that he knew that "something was behind things, something deeply hidden." He was determined to find this hidden truth, a search that would last him 76 years.

Sometimes a great moment of science is a new planet, a flying plane, or a new vaccine. But sometimes an important moment is a fifty cent compass given to a fertile mind in search of his way.

94. Jeffries and Blanchard Cross Channel by Balloon January 7, 1785

On January 7, 1785, two men, American Doctor John Jeffries and Frenchman Jean-Pierre Blanchard attempted to be the first balloonists to cross the English Channel. From Dover, England to Calais, France is about 21 miles. About half way across the Channel, their hydrogen balloon began to lose altitude at an alarming rate.

Panic stricken the two began to jettison anything in the balloon to make it lighter. This succeeded in slowing their descent, but it was clear that they would not make the French landing. With nothing else to dispose of the two quickly disrobed and tossed their clothes over the side as well. They arrived safely in France scantily clad but with their wits still in tact. A jubilant crowd paid little attention and treated them to the heroes welcome they deserved.

95. Vineland Map
Libby Pioneers Carbon -14 Dating - 1949

Guessing something's age is always a risky proposition. Sometimes however, it is imperative. This was certainly the case in 1957 when a Viking Map surfaced, called the "Vinland map" purportedly showing a continent west of Greenland and dated at 1433. This would predate Columbus' voyage and discovery of America by fifty nine years!

In 1949 scientist Willard Libby (1908-1980) had found an ingenious method of determining age. It was called carbon-14 dating.

Carbon-14 is a radioactive substance that makes up about 1 millionth of the total carbon in the atmosphere. While plants and animals are living they ingest carbon-14. The carbon-14 gets absorbed from the atmosphere during photosynthesis and animals eat these plants or they breathe it. Once an organism dies however, they no longer eat or perform photosynthesis. As the organism deteriorates, this carbon-14 slowly begins to decay into nitrogen according to a mathematical formula known as half-life. By comparing the amount of carbon-14 left in a specimen to a table showing the known amount of carbon-14 in an average living sample, one can estimate its age.

The Vinland map, dated at 1433 AD, showed America fifty nine years before Columbus ever set sail!

The carbon-14 verdict on the Vinland map found that it was a fraud. The parchment on which it is written indeed predates Columbus, but the ink is of recent origin, perhaps 1900's. It was also suspicious because unlike today, the Vikings were not inclined to make maps like their European counterparts. This is not to say that the Vikings were not here long before Columbus. Archeological digs and DNA evidence in Newfoundland and other areas of Canada confirm their presence prior to Columbus.

Today Libby's Carbon-14 dating is considered the standard for determining age. Libby's methodology revolutionized the fields of geology, anthropology, archeology and geophysics. It allowed scientists to fix a time to an object with some certainty so that they can breathe carbon-14 a little easier.

96. Wireless Power
Nikola Tesla Pioneers Wireless Power - 1893

Nikola Tesla (1856-1943) holds a wireless light bulb lit by high frequency waves

Imagine a city block with no telephone poles and no power lines. Energy for light bulbs and refrigerators fished out from the air as easily as television and radio waves. This was Nikola Tesla's dream in 1893.

Tesla stood in the center of the room holding what appeared to be a very large light bulb, although it didn't have any wires attached to it nor batteries. He shut off the lights and the room grew dark. Tesla flicked on the Tesla coil, one that he had invented, and a hum filled the room. The voltages induced from the coil climbed higher and higher as the smell of ozone gas crept into the room.

Suddenly, amazingly the wireless bulb began to come to life. "How can this be?" asked one guest. "Surely he must have hidden wires up his sleeve," asked another, "Or at least some batteries?"

Tesla responded, "What you are witnessing is a new form of power, wireless power. It is created by very high ultra frequency radio waves traveling through the air, without wires to power the bulb." Tesla smiled a bit as he too was amazed at his own genius.

" I can also run a motor with three-quarters of a horsepower right out of my hand anywhere in the room," he continued. "The secret is in the resonance, the natural frequency of the tube. If one can generate an electromagnetic wave equal to this natural resonance frequency of the bulb it will glow." Almost no one could understand his explanation but all were certainly impressed.

Tesla's wireless power concept was simple. Like radio waves or television waves of today, these waves are created in a central location, and broadcast outward to the community. Homes and businesses or anyone else that needed power would simply receive these power waves via an antenna. Tesla had gotten one bulb to light wirelessly from three or four yards away, but could he light a whole city? Tesla was going to find out.

On May 17, 1899 Tesla dropped his bag of tricks in Silver Springs, Colorado, a remote area thirsty for people and business. Here he built a large tower called a "Magnifying Transmitter." At the foot of the Colorado Mountains, Tesla erected his huge 180 foot tower with a large dome shape on top. It was capable of producing tens of millions of volts. An eerie lightning began to emanate outward from the top of the dome. One such bolt measured over 130 feet. The very skies above them seemed to be ripping apart. Little sparks of lightning were happening everywhere in the town of Silver Springs. They were coming off people's shoes, sparking to one's hands as they reached for doorknobs. Passing dogs had to think twice before approaching too close to a nearby fire hydrant. Finally one night the entire town went black. Tesla had grounded his transmitter to the city's plumbing. The entire town was without power and the local power station caught fire.

The Magnifying Transformer in Colorado, Springs

That was it. The small town of Silver Springs needed people, but not Nikola Tesla. He was forced to dismantle his plant and returned to New York City in 1899. The residents would just assume that Tesla take the sign he posted at the gate, "Stay out: Great Danger," and wear it on his back. Silver Springs wanted genius but with some degree of restraint – a word never associated with Tesla.

Tesla had stayed in Colorado Springs for nine months and had given birth to tremendous ideas even if the townspeople did not agree. Having been scientist non-gratis in Colorado, Tesla tried to set up shop on Long Island, close to New York at a place called Wardenclyffe. Unfortunately his new plant was never finished due to financial difficulties. His main financiers J.P. Morgan, George Westinghouse and others had been misled into thinking he was working on a large radio transmitter. When Tesla explained the true nature of his work, wireless power, Westinghouse and others pulled the plug on him.

While at Silver Springs he was able to light hundreds of lamps for as far as 20 miles away, but unlike television or radio waves, the power dissipated quickly with distance. Wireless power remains a tantalizing idea even today to a world that could not march to the frequency of Tesla's drumbeat.

97. Tunnel Vision
The English Channel Tunnel
Opens
May 6, 1994

Spanning the twenty-four miles that separates England and France is a treacherous body of water known as the English Channel. Over the centuries it has served as a sort of moat to castle England to protect it from would be invaders. It was therefore no small engineering or political feat when an underwater tunnel, nicknamed the "Chunnel" was opened on May 6, 1994. For the first time in over 13,000 years there was a land bridge between England and Europe.

As early as 1802 Napoleon envisioned building a Channel tunnel. When Napoleon could not defeat the English navy, he proposed building it complete with long chimneys that would protrude above the water for ventilation. It was to be big enough to carry horse drawn carriages and of course French soldiers. Envisioning it and building it are two different things and the project was abandoned.

The tunnel completed in 1994 was nothing short of a modern engineering miracle. Construction began in 1988. It took several years, 13,000 workers and the removal of enough dirt to form a small island. It could fill 68 football fields. Special boring equipment was created with a head of twenty-five yards across and a tail of over 200 yards. The French began digging on their side of the Channel, while the British began on theirs. The biggest fear is that the two tunnels would not meet in the middle. Amazingly, with only incredible surveying, without the aid of GPS, they came within 14 inches of one another after each dug underground for over ten miles.

There are three tunnels, two of the tunnels were for train service, one each way, and a central third tunnel for maintenance and emergencies. From terminal to terminal is thirty-one miles stretching from Calais, France to Folkestone, England. Twenty-four miles of the tunnel is underwater buried at an average depth of 150 feet beneath the sea bed. Over 7 million people ride these rails yearly.

What's next? Ambitious engineers and visionaries want to build a

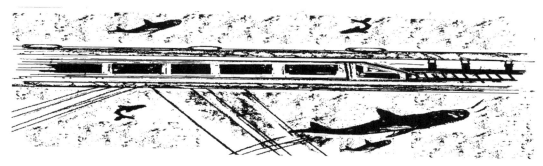

The Transatlantic Tunnel would span the 3,610 miles from New York to Paris. Traveling at speeds in excess of 5,000 mph it would take less than one hour to make the crossing.

Transatlantic Tunnel connecting Paris to New York, a distance of 3,610 miles. Using the technology used at drive up bank tellers, air will be sucked out of a long tube to reduce friction. A super high speed train would reach speeds of up to 5,000 miles per hour and negotiate the trip from New York to Paris in less than an hour. Such a tunnel's cost would be exorbitant, estimated at about a billion dollars a mile and it would take about 300 years to complete. Just as engineers thought the Channel tunnel to be impossible, proponents of the transatlantic tunnel argue that despite its enormous costs and challenges, all that is required is for mankind to think creatively and use a little "tunnel vision."

98. Bell's Other Invention
March 10, 1876

We all know that Alexander Graham Bell invented the telephone, but did you know that Bell had other inventions as well? One of these was the metal detector which did not have a great beginning.

On July 2, 1881 President Garfield was shot in a train station by a frustrated office seeker. Being a very large man, the doctors could not locate the bullet in his body to remove it. Bell, being first and foremost a patriot, rushed to Washington D.C. with a newly invented device, a metal detector, to find the offending bullet. For three days he searched the President's body but no matter where he Bell placed the metal detector near Garfield's body it beeped. A confused and dejected Bell left Washington not knowing what had gone wrong. Bell returned several times but to no avail. Eventually, Garfield succumbed to his injuries and died after seventy-nine days.

What was the source of the metal detector malfunction? It seems Garfield, being a large man, was on a new type of mattress filled with metal springs! No matter where he placed the metal detector, these springs would cause it to buzz.

While Alexander Graham Bell's first invention, the telephone, was blessed with good timing and a stroke of luck another of his inventions, his metal detector, was not. But metal detectors are very popular with those scanning the beaches for hidden treasure. For all of those who have discovered loose change or even a gold doubloon or a lost wedding ring, the invention of the metal detector was indeed a great moment of science.

99. Serendipity
Young Edison Rescues James MacKenzie
1862

Young Thomas Edison was not the best of students. "Addled" is how one teacher described him. Always into mischief, inattentive, they sent him permanently home from school not to return at age 12. Luckily for Edison, his mother was an ex-school teacher who home schooled young Thomas in many subjects and in particular science. Always resourceful, at age 15 Edison began selling newspapers and candy on the railroad that ran from Port Huron to Detroit.

One day, at the Mount Clemens station Edison saw 3 year old Jimmy MacKenzie playing on the railroad tracks in front of a rolling train car. Edison jumped into action and pulled the young MacKenzie to safety. It turns out that young MacKenzie was the son of the station manager. A grateful Mr. MacKenzie took Edison under his wing and taught him telegraphy. Telegraphy in 1862 was extremely marketable. With this skill Edison was able to leave Port Huron, and find employment in various cities until eventually he wound up in New York in 1869.

It was this skill that led to his first successful invention, a stock ticker, which was sold to the Gold and Stock Telegraph Company. This $40,000 allowed Edison to start up his first lab in Newark, NJ. Here he invented several other devices to improve the telegraph and Edison was on his way. Whenever he ran short of cash, Edison would return to the world of telegraphy and invent something new.

It was fortunate for James MacKenzine that Thomas Edison was there in 1862 and a great moment of science because it was this event that led to several interconnected events and set one of America's greatest inventors on the right track.

100. Earth Rise
Dec. 24, 1968

Apollo 8 was the first manned mission to circle the moon. On December 24, 1968, the Apollo 8 astronauts, Jim Lovell, Frank Borman, and William Anders had just finished their third lunar orbit and were coming out from behind the moon.

As they emerged they beheld the earth rising over the desolate moon surface. The contrast was stark. Above the gray lifeless moon was the brightly lit blue earth topped with swirling clouds. The bottom third of the earth was covered in shadow.

"Wow! Look at that!" Borman exclaimed. He grabbed his black and white camera and snapped a picture of what came to be known as the "earth rise" photograph. Anders remarked that it looked like a "fragile Christmas tree ornament."

It was the first picture of earth seen from another world. It underscored our tenuous position in the vast sea of space and crushing darkness. Its message was clear, the earth is a tiny lifeboat of life, don't screw it up! We need to be better caretakers of our tiny planet. The picture was even more dramatic, since the earth is seven times bigger in the lunar sky than the moon is in ours. Of the 700 photographs taken by Apollo 8, none was more important. I close the same way that the crew of Apollo 8 did "Goodnight, good luck, a Merry Christmas and God bless all of you – all of you on the good earth." As the adage goes, one picture, particularly this one, is worth 100 great moments of science.

Astronaut Anders said of the earth that it looked like "a fragile Christmas tree ornament."

BIBLIOGRAPHY

Adams, Cecil. *The Straight Dope.* New York: Ballantine Books, 1998.
Aldrin, Buzz. *Reaching for the Moon.* Harper Collins, 2005.
Balchin, Jon. *Science-100 Essential Scientists.* New York: Enchanted Lion Books, 2003.
Calaprice, Alice. *The Expanded Qutoable Einstein.* Princeton, NJ: Princeton University Press, 2000.
Casper, Barry and Richard Noer. *Revolutions in Physics.* New York: W.W. Norton & Company, Inc. 1972.
Curtis, Robert H. *Triumph Over Pain.* New York: David McKay Company, 1972.
Daley, Kenneth. *Aerospace Science: The Science of Flight.* Maxwell Airforce Base, Alabama 1993.
Fenton, Mathew McCann. *Time: Great Inventions- Geniuses and Gizmos: Innovation in Our Time.* New York: Time Inc., 2003.
Fermi, Laura. *Galileo.* New York: Basic Books, 1961.
Glassman, Gary. *Transistorized.* Boston: PBS VHS. 2000.
Hoddeson, Lillian. *Crystal Fire.* New York: WW Norton. 1997.
Jermanok, Jim. *The Wright Stuff.* Boston: American Experience, NOVA GBH, 2009.
Jones, Eric. *"Apollo Lunar Surface Journal"* NASA, 1995.
Kaku, Michio. *Physics of The Impossible.* New York: Anchor Books, 2008.
Kruff, Paul. *Microbe Hunters.* Orlando, Fl: Harcourt-Brace, 1926.
Koestler, Arthur. *The Sleepwalkers.* New York: McMillan 1959.
Lindbergh, Charles. *We.* New York: G.P. Putnam's Sons, 1927.
Lindbergh, Charles. *The Spirit of St. Louis.* New York: Charles Scribner's Sons, Inc. 1953.
Miller, Kenneth. *The Life Millennium. The100 Most Important Events & People of the Past 1,000 Years.* New York: Life Books Time Inc. 1998.
Platt, Richard. *Eureka! Great Inventions and How They Happened.* Boston: Kingfisher Publications Plc, 2003.
Sagan, Carl. *Cosmos.* New York: Ballentine Books, 1980.
Sagan, Carl. *The Cosmic Connection.* New York: Anchor Books, Doubleday, 1973.
Stefanovik-Ravasi, Slavoljib. *Tesla: Master of Lightning.* Boston: American Experience, NOVA, WGBH Boston, 2009.
Streissguth, Thom. *Rocket Man.* Minneapolis: Lerner Publishing, RCarolrhoda Books, 1995.

Wilson, Jim. http:\ www.NASA.gov
Wolfe, Thomas. *The Right Stuff.* New York: Discover Books, 1983.
Wood, Laura. *Louis Pasteur.* New York: Julian Messner, Inc. 1984.
Zitzewitz, Paul. *Physics: Principles and Problems.* Westerville, Ohio: Glencoe Macmillan/McGraw-Hill, 1992

Footnotes by Chapter

1. Kennedy, President. Speech given at Rice University, Sept. 12, 1962.
   Aldrin, Buzz. *Reaching for the Moon.* HarperCollins, 2005.
   Jones, Eric. *"Apollo Lunar Surface Journal"* NASA, 1995.
   Wolfe, Thomas. *The Right Stuff.* New York: Discover Books, 1983. p49.
2. Jermanok, Jim. *The Wright Stuff.* Boston DVD: American Experience, NOVA WGBH, 2009.
3. Curtis, Robert H. *Triumph Over Pain.* New York: David McKay Company, 1972. p61.
4. Pope, Alexander. *The Poems of Alexander Pope.* 1709. New Haven, CT: Yale University Press, 1970. p117.
   Squire, J.C. *Poems In One Volume.* London: William Heinemann LTD, 1926, page 2.
   NOVA, *Einstein's Big Idea* (2005), PBS
5. Clinton, William. Press Release from White House on June 26, 2000. http://www.genome.gov/10001356
   Franklin, Ben. France, Nov, 21, 1783 Scientific Urban Legends. http://www.Ihup.edu/~dsimanek/sciurban.htm
10. Oppenheimer, Robert J. Interview about the Trinity explosion, first broadcast as part of the television documentary *The Decision to Drop the Bomb*, produced by Fred Freed, NBC White Paper, 1965;
12. Lindbergh, Charles. *The Spirit of St. Louis.* New York: Charles Scribner's and Sons, Inc. 1953.
14. Edison, Thomas. The Franklin Institute, (2009). http://www.fi.edu/learn/sci-tech/edison-lightbulb/edison-lightbulb.php?cts=electricity
15. Farnsworth, Philo. Brigham Young High School Website. http://www.byhigh.org/History/Farnsworth/PhiloT1924.html
17. Galilei, Galileo. *Galile: Astronomer and Physicist.* Barnes and Noble, NY. 1997. Paul Hightower.

20. Goddard, Robert. *Rocketmen.* Discovery Channel, 1994. 60 min.
21. Crichton, Michael. *Jurassic Park.* Hollywood, CA: Universal Pictures. Directed Stephen Spielberg. DVD. 1993.
24. Watson, Thomas. San Francisco, Januaryy 25, 1915. Phone call.
    www.corp.att.com/history/nethistory/transcontinental.html
    Edison, Thomas. American Experience. *The Telephone. 1997.* VHS. Kirk Simon and Karen Goodman.
    www.pbs.org/wgbh/amex/telephone/.../description.html
26. A. Rupert Hall, "Galileo nel XVIII secolo," *Rivista di filosofia,* 15 (Turin, 1979), pp. 375-78, 83.
26. John Paul II, "Address to the Pontifical Academy of Sciences," November 10, 1979. p75.
28. Calaprice, Alice. *The Expanded Qutoable Einstein.* Princeton, NJ: Princeton University Press, 2000. p76.
29. Koestler, Arthur. *The Sleepwalkers.* New York: McMillan 1959.
32. Kennedy, John. Speech given at America's Cup Sept 16, 1962. Newport, RI.
    Plato. *Timaeus.* Plato quotes Critias' account of the legend, as told to Solon by one of the Egyptian priests.
34. Einstein, Albert. Calaprice, Alice. *The ExpandedQutoable Einstein.* Princeton, NJ: Princeton University Press, 2000. p83.
36. Johnson, Lyndon. Wolfe, Thomas. *The Right Stuff.* New York: Discover Books, 1983. p49.
43. Latrobe, John. Quoted by Richard Reinhart in *Working on the Railroad, 1970.* Lecture at Maryland Institute, March 23, 1868.
45. Sagan, Carl. *Cosmos.* New York: Ballentine Books, 1980. Episode 8, Journeys in Space and Time, page 161.
48. America's Story. Library of Congress: You're your States.
    www.americaslibrary.gov/cgi-bin/page.cgi/es/sd/mount_1
49. Anderson, William. thomas.loc.gov › THOMAS Home › Congressional Record. Recognizing The 50th Anniversary Of the Crossing of the North Pole By the USS Nautilis" -- (House of Representatives - July 14, 2008)
50. Adams, Cecil. *The Straight Dope.* June 6, 2006.
    http://www.straightdope.com/columns/read/2656/who-was-the-first-to-reach-the-north-pole
53. Galileo. Casper, Barry and Richard Noer. *Revolutions in Physics.* New York: W.W. Norton & Company, Inc. 1972.

69. Khrushchev, Nikita. Speech at the plenum of the Central Committee of the CPSU. Recounted by Colonel Petrov, Interview with Russia's Interfax April 12, 1964.

Photo Credits

Wright Brothers: National Geographic Image Collection, Goddard: Robert Hutchkins Goddard Library, Clark University. Tesla: *Tesla: Master of Lightning.* Boston: American Experience, NOVA, WGBH Boston, 2009